Praise for *Volume Control*

"[*Volume Control*] is the best primer I've ever read on sound and hearing, and full of advice for people of any age to consider. . . . [Owen] gives us a wonderful insight into the world of the hard of hearing and deaf."

—*The Wall Street Journal*

"Informative and entertaining . . . In clear, appealing prose, Owen explains how loud sounds—machinery, live music, etc.—can leave people no longer noticing smoke alarms, sirens, gunshots, and backup signals. . . . He makes earwax interesting. . . . The book brims with useful advice."

—*Kirkus Reviews* (starred review)

"Owen's writing and thinking about the nature of ears, sounds, and communication are lively. . . . *Volume Control* will remain relevant for decades to come."

—*Pacific Standard*

"Timely and informative . . . This well-researched and accessible introduction to the complicated subject of hearing loss is highly recommended for all science readers, not just those experiencing hearing impairments."

—*Library Journal* (starred review)

"Accessible and surprisingly entertaining . . . This work addresses an important issue for the growing pool of aging baby boomers."

—*Booklist*

"Owen, a *New Yorker* staff writer, wrestles with the complexities of the human ear in this informative [and] illuminating account of human hearing."

—*Publishers Weekly*

T0036627

"In *Volume Control*, David Owen brings his superb skills as a reporter and storyteller to the increasingly urgent issue of hearing loss. The baby boomers are aging—and so are their ears. Fortunately, and probably because of this demographic trend, both science and commerce are at last paying attention to this invisible but epidemic problem. Owen is an erudite and entertaining guide not only to the new technologies that make hearing aids better and more affordable, but to the myriad byways and curiosities he encounters in his research."

—Katherine Bouton, author of *Smart Hearing* and *Shouting Won't Help*

"David Owen aptly addresses the medical, emotional, and social aspects of hearing loss, along with some surprising revelations about technology and hearing aids. He presents the latest information in a way that makes you want to keep reading."   —Barbara Kelley, Executive Director, Hearing Loss Association of America

"As this book makes clear, most of us will encounter hearing loss at some point in our lives; we all stand to gain from reading *Volume Control* for practical reasons alone. But David Owen brims with a curiosity that's beautifully matched by his journalistic alacrity. How many times I beamed with sheer delight simply to follow the author down one fascinating path after another."   —Leah Hager Cohen, author of *Strangers and Cousins*

"A wide-ranging exploration of our vital sense of hearing, and the consequences when it wanes. Owen makes accessible not only the fascinating biology of hearing, but the complexities of remedying its loss."

—Jerome Groopman, Dina and Raphael Recanati Professor of Medicine at Harvard Medical School and author of *The Anatomy of Hope*

ALSO BY DAVID OWEN

Where the Water Goes

The Conundrum

Green Metropolis

Sheetrock & Shellac

Copies in Seconds

The First National Bank of Dad

Hit & Hope

The Chosen One

The Making of the Masters

Around the House
(also published as *Life Under a Leaky Roof*)

Lure of the Links
(coeditor)

My Usual Game

The Walls Around Us

The Man Who Invented Saturday Morning

None of the Above

High School

VOLUME
CONTROL

Hearing in a Deafening World

David Owen

RIVERHEAD BOOKS · NEW YORK

RIVERHEAD BOOKS
An imprint of Penguin Random House LLC
penguinrandomhouse.com

Copyright © 2019 by David Owen
Penguin supports copyright. Copyright fuels creativity, encourages diverse voices,
promotes free speech, and creates a vibrant culture. Thank you for buying an authorized
edition of this book and for complying with copyright laws by not reproducing, scanning,
or distributing any part of it in any form without permission. You are supporting writers
and allowing Penguin to continue to publish books for every reader.

Portions of this work have been previously published, in slightly different form,
as "High-Tech Hope for the Hard of Hearing" (March 27, 2017) and "Is Noise Pollution
the Next Big Public-Health Crisis?" (May 6, 2019), in *The New Yorker* (newyorker.com).

Riverhead and the R colophon are registered trademarks of
Penguin Random House LLC.

The Library of Congress has catalogued the Riverhead hardcover edition as follows:

Names: Owen, David, 1955– author.
Title: Volume control : hearing in a deafening world / David Owen.
Description: New York : Riverhead Books, 2019. |
Includes bibliographical references and index.
Identifiers: LCCN 2019000860 (print) | LCCN 2019002997 (ebook) |
ISBN 9780525534242 (ebook) | ISBN 9780525534228 (hardcover)
Subjects: LCSH: Deafness. | Hearing aids.
Classification: LCC RF290 (ebook) | LCC RF290 .O94 2019 (print) | DDC 617.8—dc23
LC record available at https://lccn.loc.gov/2019000860

First Riverhead hardcover edition: October 2019
First Riverhead trade paperback edition: October 2020
Riverhead trade paperback ISBN: 9780525534235

Printed in the United States of America
1   3   5   7   9   10   8   6   4   2

Book design and illustration by Daniel Lagin

While the author has made every effort to provide accurate internet addresses
at the time of publication, neither the publisher nor the author assumes
any responsibility for errors, or for changes that occur after publication.
Further, the publisher does not have any control over and does not assume any
responsibility for author or third-party websites or their content.

For Ann

Contents

VOLUME CONTROL

One

PARDON?

When my mother's mother was in her early twenties, a century ago, a suitor took her duck hunting in a rowboat on a lake near Austin, Texas, where she grew up. He steadied his shotgun by resting the barrel on her right shoulder—she was sitting in the bow—and when he fired he not only missed the duck but also permanently damaged her hearing, especially on that side. The loss became more severe as she got older, and by the time I was in college she was having serious trouble with telephones. (*"I'm glad it's not raining!"* I shouted, for the third or fourth time, while my roommates snickered.) Her deafness probably contributed to one of her many eccentricities: ending phone conversations by suddenly hanging up.

I'm a grandparent myself now, and I know lots of people with hearing problems. A guy I sometimes play golf with came close to making a hole in one, then complained that no one in our foursome had complimented him on his shot—even though, a moment before, all three of us had complimented him on his shot. (We were walking behind him.) My parents-in-law, like many older people, have a hard time ignoring a ringing telephone but also a hard time hearing what callers are saying; they have turned up the volume on their kitchen telephone so high that even if you're in another room you can't help but eavesdrop. The man who cuts my wife's hair has begun wearing two hearing aids, to compensate for damage that he attributes to years of exposure to professional-quality blow-dryers. My sister has hearing aids, too. She traces her problem to repeatedly listening at maximum volume to Anne's Angry and Bitter Breakup Song Playlist, which she created while going through a divorce. I know several people who seem to be hard of hearing but could probably be described more accurately as hard of listening—a condition that often coexists with deafness, or transitions into it, and makes it worse. One of my wife's grandfathers lost most of his hearing in old age, and another relative said of him, "He never did listen, and now he can't hear."

My own ears ring all the time—a condition called tinnitus. I blame the Chinese, because the ringing started, in 2006, at around the time I was recovering from a monthlong cold that I'd contracted while breathing the filthy air in Beijing, and whose symptoms were made worse by changes in cabin pressure during the long flight home. Tinnitus is usually accompanied by hearing loss. It's said to affect forty-five million Americans, including a surprising number of people in their teens and twenties and thirties—although so many of the people I talked to while working on this book told me they have it that I wouldn't be surprised if the real number is higher. The ringing in my ears is constant, high-pitched, and fairly loud, but I'm usually able to ignore it unless I'm lying awake in bed or, as I discovered recently, writing about tinnitus.

The National Center for Health Statistics has estimated that thirty-seven million Americans have lost some hearing. According to the National Academy of Sciences, hearing loss is, worldwide, the fifth leading cause of years lived with disability. The World Health Organization has estimated that by 2050 there will be a billion people with a disabling hearing loss. Two-thirds of Americans who are seventy or older have lost some hearing, according to various estimates. Hearing loss is also the second leading cause of service-connected disability claims made by military veterans (tinnitus is first). All this bad news is made worse by the fact that the ears you're born with are the only ears you get: a newborn's inner ears are fully developed and are the same size as an adult's, and, unlike taste buds and olfactory receptors, which the body constantly replenishes, the most fragile elements don't regenerate.

Hearing problems are often aggravated by the human tendency to do nothing and hope for the best, usually while pretending that everything is fine. This is the way we treat many health problems, although it's not the way we typically treat threats to our other senses. People who need glasses almost always get them, and, as Lauren Dragan wrote on the website Wirecutter in 2018, "If someone told you that wearing certain jeans too often might trigger permanent leg numbness, or overuse of a hot sauce would cause you to lose your ability to taste sweets, you'd pay attention." Yet people who notice trouble with their ears wait more than ten years, on average, before doing anything other than saying "Huh?," turning up the TV, and asking other people to speak up. I heard a joke about a man who was worried his wife was going deaf. He told his doctor, who suggested a simple test. When the man got home, he stood at the door of the kitchen, where his wife was at the stove, and asked, "Honey, what's for dinner?" She didn't respond, so he moved closer and asked again. She still didn't respond, so he stood directly behind her and asked one more time. She turned around and snapped, "For the third time, chicken!"

ONE DAY WHEN I WAS SEVEN OR EIGHT, I drove myself halfway crazy by staring at the house across the street and trying to figure out what seeing *is*. It leaves no trace; you can't feel or taste it; it does something to you but you can't put into words what that something is; how do you know for sure that you are doing it? Hearing is at least as hard to comprehend. For a previous book of mine, I interviewed engineers and scientists who'd been involved in the development of the Xerox machine. One of them, a physicist who'd received 155 patents during his years at the company, beginning in 1952, said, "The more you understand about xerography, the more you are amazed that it works." Scientists who study hearing often feel the same way. One of them told me, "If you stop and think about how hearing works, it seems insane." The principal components of the auditory system are coiled inside a spiraling fluid-filled chamber about the size of a pea, yet a person whose ears are fully functional can hear vibrations so faint that they displace the air molecules inside their ear canals by distances measured in trillionths of a meter. I had a couple of long conversations with a prominent hearing researcher, and, at one point, while he was using a diagram on the wall of his office to explain the still somewhat puzzling functions of two different kinds of auditory nerve fibers, the whole thing suddenly seemed so fantastic that I worried that if I learned more I might cause my own ears to stop working—like the tightrope walker who falls the moment he looks down.

Yet those of us who can hear are often extraordinarily reckless with this extraordinary gift. The greatest modern threat to hearing is excessively loud sound. Ears evolved in an acoustic environment that was nothing like the one we live in today. Thunderstorms, gales, waterfalls, ocean waves, erupting volcanoes, howling animals, screaming enemies: during most of human history, few of the world's noises would have been either loud enough or sustained enough to cause permanent hearing problems. Deafness was by no means unknown, since the pre-noise era was also the pre-antibiotics era, and infections of many kinds left eardrums inflamed or in tatters, or filled middle ears with pus, or destroyed the delicate sensors deep within the inner ear. Ears have also always been vulnerable to mischief and accidents and fighting and warfare and genetic glitches. But our ability to deafen ourselves with ordinary daily activities has never been greater than it is now. Grown-ups often assume that the population segment in the most danger is teenagers who listen to loud music through earbuds, but almost all of us routinely expose ourselves to sound levels that are potentially damaging. Although we are generally more aware of the dangers of noise than people were in the past, and are therefore more likely to take steps to protect ourselves, the world is louder as well—so much so that, for virtually everyone, completely avoiding damage is impossible. Hardly anyone makes it to retirement age with their ears in anything like their original condition.

One reason for our recklessness is that most of us underestimate the importance of hearing to our well-being. An occasional activity for my friends and me, when we were lads, was to consider whether we'd prefer to be frozen to death or burned, hanged or guillotined, shot by a firing squad or drowned—a classic sleepover thought problem. We also debated whether we'd rather be deaf or blind, although that debate seldom lasted very long, because, like most people who are able to do both, we assumed that not being able to hear would be a minor infirmity compared with not being able to see. My grandmother got deafer and deafer but still lived what appeared to me to be a mostly normal grandmother life. (Eyesight eventually became an issue for her, too—although, when she was in her late eighties and had to renew her driver's license, "the nice man at the triple-A" helped her identify the symbols in the eye test, which she couldn't make out on her own.) But my friends and I didn't actually know enough about either blindness or deafness to make an intelligent choice.

When Helen Keller was nineteen months old, in 1882, she contracted what her doctor called "an acute congestion of the stomach and the brain"—a disease that's now believed to have been either bacterial meningitis or scarlet fever—and the infection destroyed both her eyesight and her hearing. When she was twenty, and had therefore lived with both disabilities for almost two decades, she didn't hesitate in her own choice. "The problems of deafness are deeper and more complex, if not more important, than those of blindness," she wrote in a letter to James Kerr Love, a pioneering Scottish physician who worked with the deaf and was a friend. "Deafness is a much worse misfortune. For it means the loss of the most vital stimulus: the sound of the voice that brings language, sets thoughts astir and keeps us in the intellectual company of men." The quotation most frequently attributed to Keller—"Blindness separates people from things; deafness separates people from people"—is something she probably never said, at least in those words. (It's often quoted, never sourced.) But she clearly believed essentially that. In 1955, when she was in her seventies and was asked The Question for what must have been the millionth time, she replied that, "after a lifetime in silence and darkness," she knew that "to be deaf is a greater affliction than to be blind," and added, "Hearing is the soul of knowledge and information of a high order. To be cut off from hearing is to be isolated indeed."

One evening not long ago, my wife and I went to a local lake for a picnic dinner with a dozen friends, and, because I'd been working on this book and therefore thinking obsessively about my ears, the old sleepover question came into my mind. I realized that if I were blind I wouldn't be able to see the children splashing in the water, or the sun going down over the far end of the lake, or the people sitting at our long picnic table, or the emails and text messages that I'd been checking surreptitiously under the table. But after I'd thought about it for a while I realized that, if I were deaf, I wouldn't be even a tangential participant in the evening—which had far less to do with watching the sunset than with engaging verbally with friends. I'd have been a silent lump at one end of a bench, trying to seem interested and present but having no idea what anyone was laughing about, and worrying that everyone was feeling sorry for me, if they were thinking about me at all. You can interact with a blind person for quite a while without realizing that they're blind; the same doesn't happen with someone who's deaf. I have a retired friend who, at social gatherings, almost always sits in silence and scowls at everyone else. He has a reputation for being sullen and ill-tempered, but his real problem, I now understand, is that he's both hard of hearing and too stubborn to wear hearing aids. So, now, with confidence, I make my final selections: *frozen, guillotined, firing squad,* and *blind.*

LUCKILY FOR MOST OF US, natural selection overengineered the human auditory system, so that even though civilization has become louder and louder, most people have retained a useful fraction of their original capability. What's remarkable is not that so many of us have trouble following conversations at cocktail parties but that anyone past adolescence can hear anything at all. Still, the consequences of even moderate hearing loss can be grave. People who are hard of hearing die younger, on average, than people who can hear well, and as they age they spend more money on healthcare of all kinds. One reason may be that older people often mishear instructions from their doctors, or don't hear them at all, and therefore don't do whatever it is they need to do to get better. In addition, people who are hard of hearing don't always notice smoke alarms, sirens, gunfire, backup signals, or approaching thunderstorms, and they occasionally forget to turn off the engine of their keyless car, because they can't hear it, and leave it running in their garage and die of carbon monoxide poisoning or suffer brain damage during the night. They are more likely to step in front of moving vehicles, and to drive their own vehicles into dangers that people with fully functioning ears are able to avoid.

Deafness also affects activities that don't seem closely connected to ears. People who play golf, hockey, and tennis have all been shown to depend heavily on aural feedback, and to have more than the usual amount of trouble hitting solid shots if they're prevented from hearing what they're doing. The problem in golf isn't with the immediate action, since the sound of a club striking a ball doesn't reach the ears until after the ball is gone; the problem is with subsequent strokes, because during a round the parts of the brain that control the swing continually recalibrate themselves, based partly on information they receive from the ears. Arnold Palmer began losing hearing when he was in his early forties and played with hearing aids for almost half his life. "Without my aids, I lose all feel for what I want to do," he once told Peter Morrice, of *Golf Digest*, which in 2005 conducted an experiment with hearing golfers in conjunction with an independent testing laboratory. Among the findings: hearing subjects hit significantly worse shots to a par-three green while wearing sound-muffling earmuffs, and they were worse at judging how far they'd hit their putts. Liam Maguire, a Canadian hockey analyst, tried a version of the *Golf Digest* experiment with hockey players, and concluded, "You just can't handle the puck if you're not able to hear it hitting the stick. It's amazing how much hearing plays into these capabilities." Even walking can be affected by hearing loss. One reason is that the auditory and vestibular systems are physically connected, so that if there's something wrong with your hearing there's often also something wrong with your sense of balance. Another reason is that people who can't hear their footsteps are less aware of what their feet are doing, and they stumble more often.

Such challenges can be overcome. Lee Duck-hee, a South Korean tennis professional now in his early twenties, had a highly successful junior competitive career despite having been born profoundly deaf. He succeeded in part by teaching himself to see things that other players hear. Paige Stringer, who is also deaf, played on the tennis team of the University of Washington and later founded the Global Foundation For Children With Hearing Loss. She told Ben Rothenberg, of the *New York Times*, "People who were born deaf or hard of hearing may have a stronger sense of intuition in general, and tend to see subtle clues in a person's face or body language better than people with normal hearing. They are more visual, because when one sense is compromised, other senses are heightened to compensate. If my hypothesis is correct, people who are deaf or hard of hearing may have an advantage in tennis because they can pick up visual cues faster and better as to their opponent's plans, and may have better reflexes because they see things sooner." Still, playing tennis at a high level without functioning ears is tremendously difficult. Lee Duck-hee is now a professional. In the spring of 2017, he was ranked 130th in the world—the highest ranking ever for a deaf tennis player. Since then, though, he's dropped significantly.

The inability to hear well is fatiguing: straining to make out what people are saying, or relying on other senses to compensate, consumes mental resources that could be put to other uses, and, largely for that reason, deafness can cause or contribute to social isolation and cognitive decline, both of which make getting older, which is itself associated with hearing loss, seem worse than it does already. Peggy Ellertsen—a retired speech-language pathologist who wears a hearing aid on her left side and has a cochlear implant on the right, and about whom I'll have more to say in chapter ten—told me that the last years of her mother's life had been profoundly affected by deafness. "It was so painful," she said. "Her isolation. Her depression. Her exhaustion. It just consumed her life." Ellertsen gave an interview to a hearing-loss website in 2017, in which she said that her mother "had become very dependent on my father, and she experienced chronic exhaustion posed by the listening demands of her condition." She added, "I think that much of her withdrawal was because she just was so overwhelmed by the loss of ability to easily and spontaneously participate in conversation—and by the stigma that results from that." Her mother's cause of death was listed officially as Alzheimer's disease, but the real cause, Ellertsen told me, was hearing loss.

AS MY GRANDMOTHER'S HEARING GOT WORSE, she visited a succession of ear doctors and audiologists, and she once traveled to St. Louis to see a famous specialist. They all told her the same thing: that no operation or therapy could undo the damage the shotgun had done. Many of my memories of her include her hearing aid, which was huge and was the color known to crayon users of that era as "flesh." I can easily picture her fumbling with a new battery, and fiddling with the volume control, and trying to find a comfortable position for the stem of her reading glasses, and rearranging her crisply hair-sprayed white hair once she'd gotten everything more or less the way she wanted it. I can also remember a sound that she herself often couldn't hear: the piercing squeal of feedback, which occurred when her hearing aid's microphone picked up the amplified sound from its own speaker.

There's still no cure for the kind of hearing loss my grandmother had, but researchers at institutions all over the world are making progress, on numerous fronts, in the search for one. One of the advantages of being a member of the baby boom generation is that, because we're so numerous, capitalism almost magically anticipates and addresses our evolving needs. It provided station wagons when we were Little Leaguers, muscle cars when we got our driver's licenses, thrifty imports when we were struggling young grown-ups, minivans and SUVs when we became parents, overpriced two-seaters when we had our midlife crises, Priuses as we approached retirement, and Uber and Lyft when we realized that we might not always be willing or able to drive ourselves around.

And, now that we're losing our hearing, the market has turned its attention to our ears, with which we've been as reckless as we have been with the earth's once seemingly inexhaustible store of natural resources. Our hearing problems have multiplied as we've gotten older—but solutions and remedies and protections and palliatives have multiplied, too. Hearing aids are improving and becoming more versatile, and the decades-old laws and business practices that have made them unaffordable for most of the people who need them are changing. Inexpensive high-tech substitutes, including apps for smartphones, are increasingly available. Relatively soon, physicians may be able to reverse losses that have always been considered hopeless. By the time we truly can't hear what our spouses and coworkers are telling us—and before our children and grandchildren have ruined their own hearing with the technological marvels that they've acquired with help from us—our deafness may be curable with a pill or an injection or an outpatient operation or a snip of a chromosome. Even tinnitus, which has defeated all efforts at eradication, may fully yield to relatively simple treatments or techniques. A scientist who has helped develop inexpensive alternatives to traditional hearing aids told me, "There is no better time in all of human history to be a person with hearing loss."

Two

OUR WORLD OF SOUND

Y ou stand at a lectern, ready to give a talk to a meeting of audiologists—licensed healthcare professionals who diagnose hearing problems and prescribe hearing aids. You lean forward and ask, "Can everyone hear me?"

As you speak, your vocal cords vibrate, and as they vibrate they jostle the air molecules inside your throat, and those air molecules jostle the air molecules inside your mouth, and the jostling spreads outward from your lips and into the air molecules that fill the conference room. It isn't the molecules themselves that travel to your listeners; it's just the commotion that began at your vocal cords, in a chain reaction of invisible fender benders. Tiny, rapid, localized fluctuations in air pressure spread from your mouth to the ears of the people sitting before you—push, pull, push, pull. We call those fluctuations sound waves.

The audiologists can indeed hear your question, and when they respond the effect is so rapid as to seem instantaneous. It isn't, though. It takes time for air-pressure pulses to move from your mouth to your audience, and it takes time for their responses to move from their mouths back to your ears. The time it takes is the speed of sound, which is determined by the combined effects of the elasticity and density of whatever medium the pulses are traveling through. And there has to be a medium: a gas, a liquid, a solid. In the air of the conference room, the speed of sound is roughly 343 meters per second, or about 767 miles per hour.

It's hard not to think of sound as a *thing*—as a beam of invisible noise particles. But it's not a thing; it's just the transmission of mechanical energy through the molecules of a physical substance. The science of acoustics is really the study of mechanical disturbances as they pulsate through stuff. If there's no stuff—as in outer space or in any other true vacuum—there can be no sound, because if there's no stuff there's nothing to vibrate, nothing to push and pull. In 2016, in an article on the website of the magazine *Acoustics Today*, Andrew Pyzdek, a PhD candidate at Penn State, described the speed of sound as "the speed of mechanical information, the speed that change travels through a material." This notion, at first, seems weird in the extreme, at least to a nonscientist, but it's actually clarifying. Imagine that you and I are sitting at opposite ends of a long wooden table. You place your hands on the table in front of you, and I strike my end with a hammer. You hear the blow, because the vibrations created by the hammer's impact travel through the air to your eardrums—at 767 miles per hour, the speed of sound through air. You also feel the blow, because the vibrations created by the hammer's impact travel through the table to your hands—at 8,859 miles per hour, the speed of sound through wood. We tend to think of the vibrations we *hear* as entirely different from the vibrations we *feel*, but they're both examples of what Pyzdek calls "mechanical information." In the case of the hammer blow, one action creates vibrations in two different media, and the vibrations travel through those media at two different speeds. One action, two disturbances, two pathways, two speeds of sound, two sensations.

Every vibration in every medium repeats at a characteristic rate—push, pull, push, pull—and that rate is called its frequency. Frequency is represented as the number of repetitions per second, or *hertz*, after Heinrich Rudolf Hertz, a nineteenth-century German physicist. Vibrations caused by distant thunder or the annoying bass line on the stereo of the car idling next to yours at the stoplight repeat very slowly, at around 30 or 40 hertz; most ordinary conversation falls between 500 and 3,000 hertz; the vibrations from a running chain saw repeat very quickly, at around 4,000 hertz. A youngish human with healthy ears can hear sounds at frequencies down to about 20 hertz and up to about 20,000 hertz, or 20 kilohertz. That's a range of ten octaves, or three octaves more than a piano. Elephants can hear vibrations far lower in frequency than we can, and porpoises can hear much higher, up to about 200,000 hertz. Dogs also can hear very high frequencies (though not at porpoise levels). Frequencies above the range of human hearing are "ultrasonic"; those below are "infrasonic." Ultrasonic and infrasonic frequencies can affect us in various ways, but we can't hear them.

Under different circumstances, our auditory systems could have evolved to handle frequency differently from the way they do—by hearing only what we now call ultrasonic frequencies, for example, or by interpreting high-frequency vibrations as low rumbles, and low-frequency vibrations as high whistles. When you and I discuss sound (voices, music, birds singing in trees, lectures to audiologists), what we're actually discussing is one small subset of all the mechanical disturbances that occur constantly in the world around us, as converted by our auditory systems into the subset of intelligible information which we call sound. Sound isn't a thing in itself. It's just what our brains make of one part of the vibrational hodgepodge that enters our ears.

Surprisingly, therefore, the answer to the old brainteaser about a tree falling in a forest—if there's no one there to hear it, does it make a sound?—is no. What we hear when a tree falls is a creation of the parts of our brain that give meaning to particular kinds of physical disturbances acting on our ears. If there were no ears in the world—if there were no creatures with organs that detect and interpret vibrations as we do—there would be no sound as we define it. The vibrations, the pulses of mechanical energy, would still exist, and we might sense them with other parts of our body, but the meaning we now give to them would not exist—as is already the case with ultrasound and infrasound, the frequencies too high or too low for us to perceive.

Or imagine that we had ears exactly like the ones we have now, but that our brains translated their nerve signals into colors, or sensations of hot and cold, or shifting combinations of basic tastes, or prickly tinglings in specific fingers and toes, or remembered images from childhood, or feelings of tranquility or anger, or something we can't imagine. What we think of as sound has as much to do with us as it does with the world.

WHEN THE AIR-PRESSURE FLUCTUATIONS initiated by your vocal cords reach the audiologists in the conference room, air molecules inside their ear canals repeatedly push against their tympanic membranes—their eardrums—which are concave circles of thin tissue an inch inside their heads. The force that the air molecules exert is called sound pressure, and, assuming that the listeners have functioning ears, the greater the sound pressure, the louder the sound that they perceive.

On the other side of each eardrum is a snug air-filled chamber, the middle ear. Inside the middle ear are three linked bones: the malleus, the incus, and the stapes. They're the smallest bones in the human body, known collectively as the auditory ossicles. Their individual names are Latin for objects they vaguely resemble: mallet, anvil, stirrup. (*Ossicle* is Latin for "tiny bone"; when people began describing the anatomy of the ear, they needed lots of synonyms for *little*.) The ossicles function the way a lever does, by translating a small force exerted over a large distance on the relatively large eardrum into a much larger force exerted over a small distance on the relatively small inner ear. The teensiest of the ossicles is the stapes, which is similar in shape to a hummingbird's wishbone but roughly half the size, maybe three millimeters across. Its curving legs rise from a flat platform of bone, the footplate, which covers the oval window, an opening into a fluid-filled chamber called the vestibule. The vestibule leads to the cochlea, a snail-shaped structure less than half the diameter of a dime. As the ossicles move, the footplate jiggles in the oval window, like a child jumping on a bed, and pushes against fluid on the other side. Cells within the cochlea convert the mechanical energy of those vibrations into electrical signals, which stimulate nerve impulses that our brains interpret as sounds. The fluctuating hydraulic pressure inside the cochlea is relieved by a second, membrane-covered opening, the round window, situated near the oval window. Push, pull, push, pull.

Each middle ear is connected to the throat, near the adenoids, by a Eustachian tube, which is a skinny conduit that equalizes air pressure on both sides of the eardrum and drains fluids that end up in the middle ear, such as the pus from an ear infection. Eustachian tubes were named for Bartolomeo Eustachio, a sixteenth-century Italian anatomist, who, in addition to being the first person to describe much of the physical structure of the auditory system, discovered the adrenal glands. Opening your Eustachian tubes is what you're trying to do when, after a sudden change in airplane cabin pressure has made you think your head is about to explode, you swallow, chew gum, or hold your nose and blow gently—and the popping or clicking you hear when you do that is the sound the tubes make as they open. Under ordinary circumstances, your Eustachian tubes function without conscious assistance from you, but when you're flying or congested they sometimes need help. Pinching your nose and exhaling gently against your closed nostrils is called the Valsalva maneuver. If that doesn't work, you can try the Frenzel maneuver, the Edmonds technique, the Lowry technique, or the Toynbee maneuver.

Ears are a Rube Goldberg machine, in other words. A creationist might argue that the complexity of the human auditory system is proof of so-called intelligent design, but it's actually the opposite, a testament to the fortuitous make-do ingenuity of natural selection. (Two of the ossicles in mammals evolved from repurposed bones in the jaws of ancient reptiles.) Still, the entire apparatus, however haphazardly assembled, is almost inconceivably sensitive. Along most of the range of humanly detectable frequencies, we can draw meaning from infinitesimal differences in air pressure on our eardrums. Right now, I can hear not only the clicking of my computer keyboard, the muffled roar of the oil-burning furnace in my basement, and the voice of my wife, who's talking on the telephone two rooms away (and around three corners and across a hallway), but also cars driving past my house on a road a hundred feet up a hill. There is no uniform, continuous transmission path between those cars and my ears. My office has windows, but they and their storm windows are all tightly closed, and a grove of tall white pines and blue spruces stands between my house and the road. Yet waves of fluctuating air pressure created by the engines and the tires of the passing cars make their way around and through the trees, and push sufficiently hard against my house and its windows to make them vibrate enough to nudge the air inside my office enough to make an impression in my brain—which is somehow able to distinguish those vibrations from the ones made by the furnace and my wife and my keyboard—and also, just now, by an airplane flying six or seven miles overhead. An MIT-trained physicist and electrical engineer told me that a very young child, with hearing undiminished by age and exposure to noise, if placed in an anechoic chamber—in which the walls, floor, and ceiling have been covered with materials that fully absorb the sound waves that strike them, making the chamber's interior quieter than any natural place on earth—would at least theoretically be able to hear the random collisions of air molecules. It seems like science fiction.

TWO EXTREMELY SMALL MUSCLES, one just a millimeter long, are attached to the auditory ossicles. Their function is presumably protective: when an ear is exposed to a very loud sound, the muscles contract, after a brief delay, thereby dampening the motion of the bones—a clenching response called the acoustic reflex. The dampening is small, and, because it's delayed, it softens only noise that follows the noise that caused it to occur. Some people can contract their auditory muscles voluntarily, and do so when they're exposed to or anticipate loud noises; occasionally, the muscles spasm on their own, for no obvious reason, creating a fluttering sensation deep within the ear.

Bats also have an acoustic reflex, but theirs works better than ours does. Bats locate prey and navigate in the dark by echolocation: they emit high-intensity sounds and then listen to and interpret the sounds that bounce back from objects nearby. Most of the sounds that bats emit are at frequencies that are too high for humans to hear—and that's a good thing for us, because if we could hear them some of them would be incredibly loud, as loud as if the bats were firing guns from their mouths. The intensity is so great that bats would be in danger of deafening themselves if they weren't able to temporarily disengage their own hearing. A few thousandths of a second before they make each vocalization, their auditory muscles contract tightly, disabling their ossicles; a few thousandths of a second later, the muscles relax, just in time to receive the returning echoes.

As always happens with natural selection, bats and their prey have been engaged in a life-or-death sensory arms race for millions of years. It's believed that hearing in moths arose specifically in response to the threat of being eaten by bats. (Not all insects can hear.) Over millions of years, moths have evolved the ability to detect sounds at ever higher frequencies, and, as they have, the frequencies of bats' vocalizations have risen, too. Some moth species have also evolved scales on their wings and a fur-like coat on their bodies; both act as "acoustic camouflage," by absorbing sound waves in the frequencies emitted by bats, thereby preventing those sound waves from bouncing back. The B-2 bomber and other "stealth" aircraft have fuselages made of materials that do something similar with radar beams.

As also always happens with natural selection, additional complex adaptations have cascaded through the part of the food chain that bats and moths occupy. Several types of moths are parasitized by mites that colonize their ears, which are situated not on their heads but on their abdomens, below their wings. The mites don't deafen their hosts, since deafness would be fatal to moths and mites—except in one species, in which the mites destroy the moth's tympanic membrane. In that species, though, the mites disable only one ear. They leave the other one unoccupied and fully functional, because, as the scientist who discovered their behavior described it, back in the 1950s, "were they to invade both sides they would leave their host unable to detect the high-pitched sounds of bats and perhaps of other predators," imperiling both moths and mites. How the mites prevent the uninhabited ear from being colonized by other mites is not completely understood, but it's known that the original colonizers create a trail of pheromones between the two ears and dispatch "scouts" to round up mites that have ventured into the wrong one. There is one moth species in which parasitic mites do destroy both tympanic membranes—but that species occupies a habitat in which bats are now extinct and in which deafness in moths is therefore not a lethal defect for either moths or mites.

For a moth, having just one functioning ear isn't a fatal handicap. Precisely locating objects by sound requires having more than one receiver, but moths don't need to do that in order to avoid bats: they defend themselves by flying erratically as soon as they detect a bat vocalization, a function for which a single ear works reasonably well. For a bat, though, losing one ear would be catastrophic. A bat's brain identifies the direction of specific sound sources largely by analyzing differences in arrival times between its two ears. An echo returning from a point to the right of a bat's line of motion will reach its right ear very slightly before it reaches its left, and the exact length of the lag helps the bat pinpoint its target.

Owls are also highly effective at doing this, and can draw location information from arrival-time differences as brief as thirty millionths of a second. And they rely on other auditory adaptations as well. Several owl species have ear openings that are positioned asymmetrically on their heads. In barn owls, the opening on the left points slightly downward, and the one on the right points slightly upward, and the two ears are tuned to different frequency ranges. An article in *The Sibley Guide to Bird Life and Behavior*, published in 2001, explains that barn owls "use the differences in the timing and pitch of sounds reaching each ear, along with their ability to memorize the noises made by various prey, to calculate the speed, position, and orientation of each strike. These adaptations allow them to capture prey in total darkness or when leaf litter or snow obscures the prey." Owls of some species can alter the configuration of the feathers that form their distinctively owlish "facial disk," creating a sound-gathering funnel. And owls' eyes, unlike our eyes, don't move in their sockets. The remarkable rotational range of an owl's neck, approaching that of the demonically possessed character played by Linda Blair in *The Exorcist*, enables it to smoothly turn its head until a sound signal is perceived by both ears simultaneously, and its eyes, therefore, are aimed directly at the source, further sharpening its ability to precisely locate prey. A few years ago, in new snow outside my back door, I found impressions made by an owl's wingtips and talons. Re-creating what had happened was easy. Leading up to the point of impact, but not beyond it, was an orderly trail of unhurried footprints made by a mouse. I could tell that the mouse had suspected nothing.

HUMANS ALSO USE ARRIVAL-TIME DIFFERENCES to localize sounds, as I demonstrated to myself recently when my wife called to me from another room while I happened to be wearing a noise-canceling earbud in just one ear. I could hear her clearly, but I had no idea where she was—in the kitchen? upstairs? in the basement? The sound seemed to be coming from everywhere, or from nowhere. This is a difficulty that my friend and semi-regular online bridge partner David Howorth deals with all the time. He lost the hearing in his left ear many years ago, and so has no ability to localize.

"There's no difference between background noise and foreground noise when you have one working ear," he told me. "It's awkward when someone shouts my name on a crowded street, because I have to look everywhere. And it's disconcerting when I'm driving and hear a siren but can't immediately see a vehicle with flashing lights. When an elevator dings, I never know which door is about to open. I tend to press the button, then walk off to a spot where I can see the entire elevator bank at once." Howorth's wife, who died a few years ago, used to serve as his "defense" when they went to dinner with other people, by sitting on his left. "If anyone else was in that seat, I'd always mention my deafness, because otherwise I would appear to be ignoring them." He recently got hearing aids intended specifically for people who have single-sided deafness. The left-ear unit has a microphone but no speaker; it transmits sound from his deaf side to the unit in his right ear. His hearing aids have enabled him to hear things he couldn't hear before, he told me, but have not given him the ability to identify the sources of sounds.

Another person I know who has single-sided deafness told me that as soon as he has visually identified the source of a sound—his wife, a radio, a siren—the sound seems to be coming from the right place, just as it did when he had two working ears. As with many of the marvels of the auditory system, this is a trick not of his ears but of his brain, which remembers enough about directionality to create the illusion of it as soon as it has gathered sufficient supporting data. The same person told me that, before he had lost all the hearing in his deaf ear, he tried a simple experiment to assess the extent of his deficit. "If a person with normal hearing puts on headphones and listens to monaurally recorded music"—music that was recorded with a single microphone—"they get what's called fusion: the identical signals from the two speakers combine into one, and the music seems to be coming from right in the middle of their head," he told me. "So I tried that. The loss in my right ear at that time was quite severe, and I wanted to see how far I would have to turn the balance knob on the amplifier to re-create the illusion that the sound was coming from the middle of my head. And the answer turned out to be zero. My brain totally compensated for the loss, and I didn't have to adjust the knob at all."

Humans with two functioning ears have at least some ability to echolocate. People who are blind sometimes develop an uncanny skill at avoiding obstacles, and they do it the way bats do, whether consciously or not, by drawing information from the surrounding sound environment. That information includes echoes resulting from sounds that they themselves have made—for example, by tapping a long white cane. And they're able to do it even though another trick of the brain is to perceive two identical sounds, including a musical note and its reverberation, as a single sound if both arrive at the ears within a narrow time window (measured in milliseconds). It's this phenomenon that prevents (most) concert halls from sounding like echo chambers, at least when they're full of people. The seats in the Oslo Opera House, in Norway, were designed with cushions that absorb sound in just the way human bodies do, so that music sounds the same no matter how full the auditorium is. Before it opened in 2008, singers in the old opera house had to adjust their voices based on whether they were rehearsing in front of echoey empty seats or performing in front of sound-absorbent humans.

Even though the brain, in effect, merges an original sound and a close-arriving reverberation, it still knows which is which—a phenomenon called the precedence effect, or the law of the first wave front—and therefore also knows which direction to turn the head in order to look toward the source of the sound. No less remarkable is the fact that those of us with two working ears hear single sounds as single sounds, rather than as separate right-ear and left-ear "auditory objects," just as signals from our two eyes merge into a single image—again, thanks to the three-pound supercomputer encased by our skull.

A FEW YEARS AGO, IN DALLAS, I went out to dinner with a large group. I was seated near one end of a long table and couldn't hear what anyone at the other end was saying. This difficulty in understanding speech against a background of noise is a nearly universal problem for people over a certain age, and the situation in which they are the most likely to notice it is when they are eating out. (I've asked many people in their fifties and older how their hearing is, and the most common response begins something like, "Well, in restaurants. . . .") Peter W. Alberti, a former chairman of the Department of Otolaryngology at the University of Toronto, explained why, in a book about occupational noise exposure: "In a crowded noisy room a young person with normal hearing can tune in and out conversations at will. This is known technically as the cocktail party effect. The brain quite automatically adjusts time of arrival and intensity differences of sound from different signal sources so that the one which is wanted passes to the cortex and all others which do not meet these criteria are suppressed by feedback loops. This requires . . . good high frequency peripheral hearing, two ears and an additional central mechanism. Even in the presence of normal bilateral peripheral hearing, the elderly lose part of the central mechanism and find it difficult to listen in crowded rooms. This is compounded if there is some hearing loss."

Many bar and restaurant managers exploit their customers' hearing problems by turning up the volume and tempo of their background music as the night goes on, both to drive away loitering oldsters and to encourage younger diners to talk less, eat faster, and drink more. Modern ideas about design—metal furniture, no carpeting, open kitchens, easy-to-clean hard surfaces—contribute to noise levels, too. (They also threaten the hearing of employees, who are exposed continuously, day after day, for hours at a time.) And all this is true even though, when customers complain about restaurants they've visited, the main thing they complain about—ahead of illegible menus printed in light brown ink on dark tan paper—is noise. SoundPrint, an iPhone app described by its creators as "basically Yelp for noise levels," enables diners to measure and share decibel readings whenever they eat out. The app's developers have determined that, at least in New York City, Chinese, Indian, and Japanese restaurants are the quietest, and Mexican restaurants the loudest. SoundPrint is a terrific idea, in my opinion, despite the paradoxical further fact that people who complain about noisy restaurants often avoid ones that seem boringly quiet: how good can they be if the diners just sit there in silence? Not surprisingly, I guess, loud restaurants are, on average, more profitable than quiet ones.

In Dallas, I was the oldest person at the table, so I assumed that my hearing problem was age-related and was limited to me. But then a young guy sitting across asked the young guy sitting next to him whether he could hear anything. "No," he said. "I'm just nodding and smiling." A little later, one of the young guys cupped his ears—turning his hands into ear trumpets, which were hearing aids in the era before hearing aids. The other young guy asked if doing that helped. He said that it helped a lot, so we all tried it—and, indeed, I found that I could now home in on particular speakers, including ones sitting at other tables. Cupping your ears can be remarkably effective if you don't mind being seen doing it. My golf buddy who didn't hear my friends and me complimenting him on his narrowly missed hole in one finally got hearing aids, a year or two later, but he told me that when he's wearing them he still has trouble hearing in movie theaters unless he also cups his ears.

When you cup your ears, you in effect enlarge each ear's pinna, or auricula, the visible external part of the ear. Many animals have large pinnae, and some can aim them at whatever they're trying to listen to. A human's pinnae are relatively small, and, although some of us can wiggle our ears very slightly if we try really, really hard and practice a lot, we can't significantly alter their shape or aim them. Our pinnae are therefore often said to be vestigial, but they actually still have a function, since they help to gather sound waves, as a funnel would, and they make it easier for us to determine which direction sounds are coming from, by gathering more sounds from in front of us than from behind. A man my father knew in grade school raced cars as an adult, and he lost one of his pinnae in a crash. He had a normal-looking ear on one side of his head, but just a round opening on the other. The crash hadn't affected his inner ear. Still, he couldn't hear quite as well as he would have been able to if he'd taken up a less dangerous hobby, and he had slightly more trouble localizing sounds.

During the quarter century between the invention of the airplane and the invention of radar, the armed forces of a number of countries employed aircraft detectors that were essentially gigantic ear trumpets. Some looked like megaphones; some looked like tubas; some were portable and were worn on the head, like a funny hat; some were enormous and were mounted on swiveling stands like the ones that support antiaircraft guns. Operators would scan the skies, listening through earpieces, until they picked up the sound of approaching airplanes. Some versions worked the way owls' heads do, with fixed sights that were aimed so that when the detected sound was at its loudest, the source would be in the operator's field of view.

How cupped hands and ear trumpets and pre-radar aircraft detectors work seems intuitively obvious, but, as with almost everything else in acoustics, the full explanation is complicated. One clue that something tricky is involved is that an ear trumpet, unlike a telescope or a pair of binoculars, works in both directions: if you place the small end of an ear trumpet next to your ear and I speak into the large end, my voice will sound louder—but if you place the large end next to your ear and I speak into the small end, my voice will sound louder, too. Turn an ear trumpet around and you have a megaphone, not a silencer; how can that be?

Much of the explanation has to do with "acoustic impedance." Sound waves move faster and more readily through some materials than through others; that is to say, they are impeded less. Sound can pass from one material to an adjacent different material, but the greater the difference in impedance between the two, the less sound will make it through. Sound moves very fast through water—much faster than it does through air—but if you submerge your head in a swimming pool, you probably won't be able to hear the voices of the people standing in the pool right next to you, with their heads above the water. The reason is that the impedance difference between the air in which those people are chatting and the water in which you've submerged yourself is so great that most of the sound waves coming from their mouths are bouncing off the surface of the pool rather than traveling down to your ears.

There's more to it than that, and there's more than one kind of acoustic impedance. Basically, though, megaphones and ear trumpets both work by reducing impedance mismatches between dissimilar materials. Piano sounding boards, violin bodies, trumpet bells, pharynxes, and what dentists call "the oral cavity" all do something similar. So do the auditory ossicles, the bones in the middle ear, which bridge the mismatch between the air on the outside of the eardrum and the fluid within the cochlea, the diminutive heart of the inner ear.

Three

THE BODY'S MICROPHONE

The cochlea occupies a small cavity in an exceptionally unyielding part of the skull, a pyramid-shaped structure known as the petrous part of the temporal bone. The petrous bone protects the cochlea and other parts of the hearing system from blows to the head. It is valuable to people who study the genomes of ancient humans and other animals because its density makes it an unusually enduring storage vault for DNA. The same density somewhat complicates the lives of surgeons, hearing scientists, and pathologists, who need power tools to get through it.

The cochlea is slightly larger than, and similar in profile to, a miniature chocolate chip. It's shaped like a spiraling shell—*cochlea* is Greek for "snail"—or like a Barbie-size serving of frozen custard. Its interior contains three fluid-filled ducts, which wind from the cochlea's base to its tapered top. The critical duct is the middle one, the smallest of the three. Running along its lower interior surface is a strip of sensory cells, the organ of Corti, named for the Italian anatomist who first noticed it, in 1851. The organ of Corti has been described as "the body's microphone." Specific parts of it resonate with vibrations of specific frequencies, like the strings in a piano, as the duct curls inward. The organ is bounded by two membranes. On the bottom is the basilar membrane, which varies in width from roughly a tenth of a millimeter, at the end closest to the oval window, to roughly half a millimeter, at the cochlea's apex. Arrayed on top of the basilar membrane is an orderly layer of cells. The most important of these are the so-called hair cells—an unfortunate term, because it makes you think of the stuff that pokes out of old men's ears. Hair cells are not hairs. A researcher who works on hearing restoration told me, "When I talk to a non-scientific audience, I have to make sure they're not getting confused about the hair on top of their head." Hair cells are microscopic and bulb-shaped, and they are topped by neat bundles of nearly submicroscopic stereocilia. The stereocilia look a little like hairs, but only a little: they are cylindrical and relatively rigid, and they're arranged in curving ranks of different heights, and they are tinier than tiny. The tallest stereocilia measure about two ten-thousandths of an inch from base to tip; the shortest measure less than four one-hundred-thousandths. They don't wave back and forth like an undersea forest of kelp.

The top layer of this sensory sandwich is the tectorial membrane, which is a fibrous gel that extends above the tops of the hair cells like the roof of a carport. It just touches the tips of the tallest stereocilia. When sound waves from the outside world enter the fluid in the cochlea, the regions of the organ of Corti corresponding to the incoming frequencies move in response. As the hair cells move, the tips of their stereocilia are pushed and pulled by the tectorial membrane, and, as they bend, what one scientist described to me as "trapdoors" in the stereocilia open, allowing ions from the surrounding fluid to enter the hair cells. These ions create electrical signals, and the electrical signals are converted into impulses in nerves that underlie the hair cells, and the impulses travel along a bundle of nerve fibers to the auditory centers in the brain. The result, when everything is working properly, is what we call sound.

None of this is easy to study. If you extracted an entire organ of Corti and rolled it into a ball, you could tee it up on the head of a pin. Each stereocilium is only about as wide as the smallest wavelength of visible light, and is therefore almost impossible to glimpse with an optical microscope. I did see some, though, on a computer monitor in the lab of David Corey, a neurobiology professor at Harvard Medical School. The image he showed me was of the top part of a mouse's outer hair cell, which measures perhaps one three-thousandth of an inch across and is structurally similar to a human hair cell. The stereocilia in the image looked a little like the curving clusters of bristles at the end of a fancy toothbrush. The image I saw had been assembled from thousands of successive electron-microscope scans, each taken after a focused beam of gallium atoms had etched away an inconceivably skinny slice from the top of the cell.

There are two types of hair cells in the cochlea, outer and inner. The outer ones, of which we're born with about twelve thousand, function as amplifiers. As the basilar membrane vibrates beneath them, the outer hair cells physically elongate and shorten, like little pumps. Their actual motions are minuscule, but those motions are sufficient to strengthen an incoming sound signal by about as much as a hearing aid would. The stereocilia of the inner hair cells, of which we're born with about three thousand, produce the electrical signals that initiate the nerve impulses that travel to the brain. "Humans can detect a sound that vibrates our stereocilia by about the diameter of an atom, or a few atoms," Corey told me. "We can also hear sounds about ten million times louder. Yet a stereocilium's entire operating range of motion is only about half of its diameter." Two other scientists, in a paper called "The Physics of Hearing," published in 2014, likened the movement of a hair-cell stereocilium to "a deflection of the Eiffel Tower's tip by the width of a *petite madeleine.*"

The machinery of hearing is so remarkably small and complex and hard to observe that scientists still don't completely understand how all of its components work. In 2018, Corey was one of a group of fourteen scientists who made a significant discovery about those pores that act like trapdoors, and they described their findings in a paper in the journal *Neuron.* A press release issued by the communications office of Harvard Medical School said that their research had "ended a 40-year-quest for the elusive identity of the sensor protein responsible for hearing and balance"—a protein that the same press release called "the hearing molecule." The story was picked up by many news outlets, and some of the resulting articles suggested that Corey and his colleagues, thanks to their discovery, had pretty much licked deafness. The protein, though important, is one of many "hearing molecules," some of which are still poorly understood. Hearing is a puzzle that scientists have been assembling, infinitesimal piece by infinitesimal piece, for centuries, and—as Corey would be the first to tell you—they're still working on it. Still, he and his colleagues were recently able to prevent deafness in a mouse that had a mutation in the protein they'd identified, using a therapy they'd devised based on their discovery. This is genuinely significant news, but don't throw away your hearing aids yet.

AS THE FREQUENCY OF SOUND WAVES rises or falls, so does the quality that we call "pitch," which is frequency as filtered by our ears and interpreted by our brains. Frequency is an objective, scientific fact. Pitch, by contrast, is subjective; it's just the name we give to the frequencies we can hear, and it's the way we arrange sounds in relation to one another. (Ultrasonic and infrasonic vibrations have frequencies, but, because we can't hear them, they don't have pitch.) One method of representing pitch involves letters of the alphabet. The note on a piano that musicians call middle C is a vibration with a frequency of just under 262 hertz; the highest C on a piano, at the far right end of the keyboard, has a frequency of almost 4,200 hertz. Before the late nineteenth century, musicians in different parts of the world tuned instruments in different ways, so that a note designated as middle C in a musical score might be played at different frequencies, and therefore given different sounds, depending on where it was being performed and what instrument it was being performed on. Nowadays, in Western classical music, the first A above middle C is a vibration with a frequency of 440 hertz; in the Baroque era its frequency ranged from less than 400 hertz to more than 460.

Although pitch is subjective, there isn't a huge degree of variation in how different humans interpret it—hence, orchestras. The relationship is neither universal nor permanent, however. My friend Charles Komanoff began playing the piano by ear at the age of five, after watching his two older sisters. He was eventually identified as having absolute pitch, which, in its most developed form, is the ability to unerringly identify or reproduce individual notes played randomly and out of musical context, and to accurately identify the pitch of other sounds, including those not made by musical instruments. Absolute pitch, which is also known as perfect pitch, is believed to occur in about one person in ten thousand. It's different from relative pitch, which is the ability to identify notes in relation to other notes and therefore to play music "by ear," in any key.

"I was somewhat of a prodigy," Komanoff told me, in an email. "I was playing several Mozart sonatas by age six, and I was able to blindfold-identify virtually any four- or five-note chord struck on the piano." He stopped playing piano at age nine, after losing some of his initial enthusiasm for classical music, but took it up again as a teenager. "My perfect pitch served me well in improvising with other musicians; in scoring big-band jazz arrangements (as a high school teacher, by ear, on popular jazz and rock tunes); and in picking out rock songs (Steely Dan, Grateful Dead, the Who) and jazz compositions (Miles, Trane) on piano."

Then, when he was in his fifties, something changed. He continued: "I began hearing new music—not 'old' pieces that I knew well, but unfamiliar music heard on the radio or in live performance—a whole note (two semitones) higher than it was actually being played. An E-major guitar chord would sound as F-sharp, for example."

Komanoff is now in his seventies, and the change in his relationship with music has been profound. "It is jarring, at a live performance, to see the guitarist or bass guitarist or keyboardist playing one thing while I hear another. It is also frustrating when I go to pick out a new song on piano or bass but the chords I hear in my head (and try to play) don't match the actual chords being played in the recording I'm trying to copy. I should add that I never developed a strong sense of relative pitch, relying instead on my 'crutch' of perfect pitch. While obviously I have a good enough sense of, say, blues progressions (I-IV-V-I), I nevertheless hear chords as absolute (F major) rather than relative (IV chord). This hinders my capacity to reprogram my listening: occasionally I can convince myself that the chord I hear as G is actually F; but then the next chord comes and I slip back into hearing it one whole note too sharp."

Komanoff's problem is not unique; it affects an unknown but significant percentage of other people who have absolute pitch, usually beginning in their fifties or sixties. (Three distinguished fellow sufferers: Gustav Mahler, Dmitri Shostakovich, and my sister-in-law Nelie McNeal.) Daniel J. Levitin—a neuroscientist, cognitive psychologist, and talented musician—told me that the cause is age-related stiffening of the hair cells. "When the hair cells stiffen, they no longer activate the same tuning curves along the basilar membrane," he said. "It's also been shown in two case studies that ingestion of the anti-seizure medication carbamazepine can cause a similar but reversible pitch shift in AP possessors." The same change presumably occurs in many non-musicians as they age but goes unnoticed by those of us who, like me, were asked by our grade school music teacher to lip-synch during music programs.

BRANCHING AWAY FROM THE COCHLEA, on the other side of the inner ear, is the vestibular system, which governs balance and spatial orientation. Its most conspicuous features are three looping fluid-filled tubes, the semicircular canals, which look like intersecting miniature teacup handles. At the base of each loop is a bulb-shaped swelling, which contains hair cells that are structurally similar to the hair cells in the cochlea. When the head rotates or tilts, the fluid inside one or another of the loops moves, too, and the stereocilia inside move in response. As they do, ions from the surrounding fluid flow into them, creating electrical signals that are then converted into nerve impulses—the same sequence of events that occurs when sound vibrations enter the cochlea, but with a different cerebral destination and a different result. The semicircular canals enable us to consciously sense movements of the head. They also control an unconscious circuit that moves our eyes when our head moves, thereby preventing light images on our retinas from jumping around if we don't keep perfectly still. The whole system works like a sort of internal Steadicam. It enables us to, for example, check our email on our phone and watch CNBC on a TV set mounted on the wall while puffing away on a treadmill or a vertical climber at the gym. Other parts of the vestibular system contain two other groupings of hair cells. Overlying their stereocilia is a gelatinous membrane that contains a large number of otoconia, which are extremely small crystals of calcium carbonate. When the head suddenly moves—as it does when an elevator we're riding in abruptly starts or stops—or as gravity continuously tugs on the otoconia, the membrane shifts position and, in doing so, exerts a force on the stereocilia below, sending more nerve impulses to the brain. It's generally believed that hearing evolved from vestibular functions, rather than the other way around, by exploiting some of the same cellular equipment. For ancient organisms, knowing which way was up most likely preceded, and directly led to, the ability to hear.

A balance problem that my *New Yorker* colleague Patty Marx had recently is benign paroxysmal positional vertigo (BPPV), which occurs when an otoconium detaches from its gelatinous matrix and becomes lodged in one of the semicircular canals. Gravity acting on the otoconium then stimulates the wrong hair cells—the ones that detect rotations of the head—and the result is a phantom sensation of turning. Typical symptoms include dizziness, certain involuntary twitches of the eyes ("nystagmus"), the head-spinning sensation that drinkers call the whirlies, and nausea. Attacks are usually brief and are almost always initiated by specific changes in head position, such as lying down to go to sleep. BPPV often goes away on its own, but it can recur, and incidents become more common with age. Marx described her symptoms to another friend, who diagnosed the condition correctly and told her, also correctly, that the way to treat it was with the Epley maneuver, a sequence of head and body movements whose purpose is to return the rogue particle to its chamber. "But don't try it on your own, because if you do it wrong you can screw yourself up permanently," the friend said. So Marx called her doctor, who told her that the Epley maneuver was indeed what she ought to do but that she didn't need to worry about trying it on her own, because even if she did it incorrectly she couldn't hurt anything. She found an instructional video on YouTube and followed the steps, under the supervision of her partner, Paul Roosin. "My problem went away almost instantly," Marx told me. The Epley maneuver works like one of those tabletop labyrinth games in which you roll a marble from one end of a maze to the other by turning knobs that tilt the maze. You keep turning your head until the wandering crystal has tumbled out of the teacup handle it tumbled into.

LOSING THE ABILITY to effortlessly orient yourself in space can be as devastating as losing your hearing or your sight. David Corey told me about a physician whose vestibular system was ravaged, in the late 1940s, by a drug he was being given for what was believed to be tuberculosis in a knee joint. Upon waking one morning, in the hospital where he was being treated, the physician briefly covered his face with a steaming washcloth before shaving—and fell down. Then, very quickly, his symptoms became dramatically worse.

"Every movement in bed now caused vertigo and nausea, even when I kept my eyes open," he wrote in the *New England Journal of Medicine*, in 1952. "If I shut my eyes the symptoms were intensified. At first, I found that by lying on my back and steadying myself by gripping the bars at the head of the bed I could be reasonably comfortable. Later, even in this position the pulse beat in my head became a perceptible motion, disturbing my equilibrium." In the byline of the article, the physician identified himself only as "J.C." He was actually John D. Crawford, who, at the time of his tuberculosis treatment, in 1948, was twenty-eight years old and a patient at a U.S. Army hospital in Germany. He had graduated from Harvard Medical School four years earlier, and since 1946 he had been serving in the same hospital as the chief of the diphtheria ward. Not long before his unfortunate drug reaction, he'd suffered a worse disaster: his wife and infant daughter were killed in a plane crash while traveling to see him.

Crawford's account of his experience is so closely observed that it's still quoted in medical literature. When he turned his head from side to side, he wrote, he had "the sensation that the room turned around me, rather than that I was turning around in the room." He likened the world as he saw it now to a movie made by someone walking down a street with the camera held snugly against their chest, so that "the street seems to career crazily in all directions, faces of approaching persons become blurred and unrecognizable and the viewer may even experience a feeling of dizziness or nausea." Crawford, initially, struggled even to read in bed. "I found that by bracing my head between two metal bars at the head of the bed I could minimize the effect of the pulse beat, which made the letters on the page jump and blur. I gradually learned to keep my place by using a finger or a pencil on the page."

Over a period of years, Crawford developed other compensatory tricks. He taught himself to walk unaided in daylight—although in darkness he sometimes had to crawl. His own motion still turned the world into a blur, so when he went for walks he began greeting anyone who approached him, in case the jittery shape looming toward him turned out to be someone he knew. If he needed to read a sign, he stopped and stood perfectly still; if he wanted to see something overhead, he stopped and, before looking up, held on to something solid. He discovered that walking was easier if he kept his eyes focused on distant objects and followed "the same type of course that is steered by a ship's gyroscopic compass, veering first slightly to the left and then overcompensating and veering equally to the right." He swam, despite warnings from his doctor, because he figured that if he became disoriented he would be able to find the surface visually—but he didn't swim at night. He switched from singles tennis to doubles, which he found that he could play while standing virtually still. He learned to identify gently rising and falling slopes underfoot by noticing changing tensions in his leg muscles. He even found one advantage: no more seasickness. He remarried, had more children, and pursued a long, distinguished career, in pediatric endocrinology, at Massachusetts General Hospital. He died in 2005, at the age of eighty-five. His success at using his remaining senses to cobble together a sort of handyman replacement for his vestibular system inspired similarly affected people to do the same—and, indeed, a number of the techniques that he devised are similar to the ones that form the basis of what's known today as vestibular rehabilitation therapy. Crawford's self-treatment also gave him a renewed appreciation for the versatility of "the human apparatus," with its "many alternate systems to accomplish its end." But still.

Four

WHEN HEARING FAILS

The damage to John Crawford's vestibular system was caused by a drug, the antibiotic streptomycin, which was discovered in 1943 and was the first effective treatment for tuberculosis. Even when streptomycin works on tuberculosis, though, it doesn't work quickly: at the time of the shaving incident, Crawford had been receiving injections for two and a half months without much improvement in his knee. Streptomycin potentially causes serious side effects, among them damage to the inner ear, and the danger increases when the treatment is prolonged. Crawford deduced the cause of his balance problem, and held his wristwatch to his ear to assure himself that he hadn't been deafened as well. And he rationalized about his inability to stand: "Perhaps loss of vestibular function *was* to be preferred to the consequences of tuberculous arthritis." Nevertheless, his doctor discontinued the injections two days later. None of his lost vestibular function ever returned.

Streptomycin was among the first modern antibiotics. It and

several of its close relatives can be ototoxic—that is, they can damage hearing—and mostly for that reason they're seldom prescribed in the United States anymore. They're still used in a number of other countries, though, because they're cheap and readily available. And in the United States they're used in cases where no other treatment exists, or where the possible damage is believed to be outweighed by the likely benefits, or where the application is topical. The Food and Drug Administration approves streptomycin for the treatment of plague; gentamicin is often administered to premature newborns as a defense against gram-negative bacterial infections, which can be lethal; tobramycin is used in treating infections that accompany cystic fibrosis; neomycin is the active ingredient in Neosporin, a widely used (and perfectly safe) nonprescription antibiotic ointment.

The chemotherapy drugs cisplatin and carboplatin are also potentially ototoxic. My son's father-in-law underwent chemotherapy twice, for bladder cancer and its recurrence, and he now wears hearing aids in both ears. Most cancer patients would accept that trade-off, although they aren't always told about it in advance. Quinine, Oxy-Contin, Vicodin, aspirin, and scores of other drugs can be ototoxic; so can a long list of compounds commonly found in industrial workplaces, among them acrylonitrile, carbon monoxide, lead, styrene, and toluene, plus any number of pesticides and solvents—all of which ought to be avoided anyway, of course, but are often present in places where they shouldn't be.

Hearing and balance losses caused by antibiotics and chemicals are called sensorineural losses because they're characterized by damage to the (sensory) hair cells and the (neural) nerve cells that carry signals to the brain. Sensorineural hearing loss has other causes as well, including advancing age, autoimmune disorders, genetic bad luck, and many viral infections, among them chicken pox, mumps, herpes, influenza, and COVID-19. I recently received an email from Susan Spiritus, who earned a master's degree in deaf education in 1968, then spent two years in Guam with her husband. "Three years earlier, sixty babies had been born deaf, as there had been a horrible epidemic of German measles on the island, and their mothers had contracted it in their third trimester of pregnancy," she wrote. German measles is a common name for the viral disease rubella. Like many people my age and older, I had rubella as a child and got over it without a problem, but for developing babies it can be catastrophic. Possible effects on children whose mothers had rubella while pregnant include cataracts, heart defects, intellectual deficits, and deafness—all good reasons to make sure that everyone receives all the vaccinations they're supposed to. "The children were three years old when I arrived, looking for a job," Spiritus continued. "I was hired immediately by the government, and before I left the island all of my students had been fitted with hearing aids, had learned to speak (orally), and had also learned sign language—which I did not teach them!" Four years after her return, Spiritus herself lost hearing on one side, as a consequence of having contracted another frequent cause of sensorineural hearing loss, Ménière's disease, whose etiology is still mysterious.

By far the biggest cause of hearing loss is overexposure to loud sound. Corey showed me another electron micrograph, from the inner ear of a mouse that had been exposed for two hours to a sound as intense as that experienced by a person using a chain saw without ear protection. The stereocilia looked like tree trunks thrown around by a tornado. Charles Liberman, a professor of otolaryngology at Harvard Medical School and the director of the Eaton-Peabody Laboratories at Massachusetts Eye and Ear, which is affiliated with Harvard, told me, "Like all mammals, we are born with a population of sensory cells. There are only about fifteen thousand in an ear, and that is many, many fewer than the rods and cones in your eye, to say nothing of the neurons in your brain. And you've got only about forty thousand nerve cells, which is an incredibly small number compared to the millions in the optic nerve. And they have to last your whole life." Almost everyone who survives to adulthood dies with many fewer functioning hair cells and nerve cells than they started out with.

HUMANS WITH HEALTHY EARS can detect sounds along a huge range of intensities. We can hear extremely faint sounds, which push on our eardrums hardly at all, and we can hear extremely loud sounds, which push on our eardrums so hard that they do permanent damage to our cochleas. The acoustic intensity of the quietest sound that a person with undiminished hearing can hear is only about a ten-trillionth as great as that of a noise that's loud enough to cause physical pain—say, that of a gun being fired near your head. Eyesight is similar. Peter W. Alberti, the former University of Toronto professor I mentioned in chapter two, likened the breadth of the human hearing range to that of the human vision range, which, he wrote, is "as wide as seeing a candle flicker on a dark night at a hundred meters to looking indirectly into a bright sun."

The human hearing range is so wide that if we represented it linearly—1, 2, 3, 4 . . . —the digits would quickly get out of hand:

imagine a sound-system amplifier with a volume knob that had billions of little markings on it. To make things more manageable, scientists and sound technicians represent the range using a logarithmic scale. The unit they typically use to represent the loudness of sound is the decibel (dB), which was first defined during the early years of the telephone, when engineers needed a convenient way to represent losses in signal strength over transmission cables. (The third syllable is a nod to Alexander Graham Bell.) Logarithms increase exponentially. The softest sound that a healthy human can hear was designated by international agreement as 0 dB; a sound with ten times that intensity—say, a few leaves rustling in the distance—is 10 dB; a sound with ten times the intensity of the distantly rustling leaves— say, what you hear when you sit quietly in a very quiet room—is 20 dB; ten times that—say, a whisper near your ear—is 30 dB; a hundred times that—say, normal conversation—is 50 dB.

Whether describing the loudness of sound in this way makes things easier or harder to understand is a matter of debate, and not just among people who majored in English. Scientists who use logarithmic scales to describe phenomena to nonscientists—as they also do, for example, when they discuss earthquake severity—often have to spend at least as much time explaining that the numbers don't mean what they seem to mean. (An earthquake that measures 7.0 on the Richter scale releases more than thirty times as much energy as one that measures 6.0.) The difficulty in understanding decibels is compounded by the fact that there is more than one system, based on different frequency "weightings" of the audible spectrum. When I emailed an acoustical engineer what I thought was a fairly simple question about decibels, he answered almost immediately, with a fairly simple explanation—but then ten minutes later he followed up with a much longer email, containing clarifications, equivocations, and links to scientific websites, and five or ten minutes after that he followed up with another.

A further complication is that the human *perception* of sound is roughly logarithmic, too—but on a different scale. An increase from 50 dB to 60 dB represents a tenfold increase in sound intensity, but we perceive it as only about a doubling in loudness—partly because our ears function in something like the way hearing aids do, by amplifying quiet sounds more than loud ones. We perceive light in a similar way. As my sound-engineer friend explained, "If I turn down the lights by 50 percent, as measured by a light meter, you'd say I'd dimmed them only slightly. And if I cut the sound intensity in half, you'd say I'd turned down the volume just a little." If our eyes and ears didn't function in this way, our heads would require wiring that's even more complicated than the wiring they have already, and making sense of the huge range of inputs we currently handle would be much harder. (Imagine an alarm clock that suddenly *sounded* a thousand times as loud as normal conversation.) Less confusing than the decibel scale might be a scale based on a gradient of familiar sounds arising close to our ears: breath, whisper, gentle rain, conversation, dishwasher, vacuum cleaner, city traffic, lawn mower, chain saw, rock concert, rifle shot, forget it.

The decibel danger line for sustained exposure is often given as 85 or 90 dB—about a billion times the sound intensity of the softest perceivable sound, or roughly lawn mower on the Owen scale, described above. A distressing number of relatively common activities lies at or above that threshold, including car-horn honking, motorcycle riding, leaf blowing, headphone listening, phone ringing, Shop-Vac-ing, milk-shake making, subway commuting, office-party going, power-tool using, and the Fourth of July. A hearing threat that scarcely existed fifty years ago but has been a nearly inescapable part of modern life ever since is listening to (and performing) loud music. Perhaps the awesomest stereo system I've ever personally experienced was a homemade one in the basement of a retired sound engineer. I sat on a couch in the sweet spot of his speaker array while he played Eric Clapton, among other favorites, and I was powerfully reminded of what we miss when we use Bluetooth earbuds and an app on our smartphone to listen to a song that someone copied from iTunes and uploaded to YouTube. And the retired engineer played his system *loud*. He said that, in his opinion, the ideal listening level for music—the level at which the awesomest songs sound their absolute awesomest—is right around the danger line, about 90 dB.

There may be a biological basis for his preference. In 1999, two scientists at the University of Manchester, in England, published a paper describing an experiment in which they'd played music at dance-club volumes to a group of student volunteers while measuring certain physiological responses. They concluded that throbbing music at 90 dB stimulated their test subjects' vestibular systems in ways that created "pleasurable sensations of self-motion" when they were sitting still, and that such sensations might "account for the compulsion to exposure to loud music." Crank the volume sufficiently, in other words, and you feel as though you're dancing when you're not. (That pleasurable effect, my own research has shown, is enhanced by alcohol and other substances, which play their own mischief with the vestibular system.) The scientists said that the responses they'd observed appeared similar to those experienced by people on carnival rides, and that a survey they'd conducted had, indeed, found a correlation between enjoying loud music and enjoying roller coasters. They wrote that their findings were consistent with the existence of a putative "rock and roll threshold"—a volume level first proposed by a British acoustical consultant, who said that rock music had to be played at high volume in order to "work."

At some decibel levels, a single brief exposure can cause instant, permanent damage, as the shotgun blast did with my grandmother. Robert Dobie, a clinical professor at the University of Texas Health Science Center in San Antonio told me that a bigger threat to hearing than loud music (at least in Texas) is recreational shooting. Another professor I spoke with, at a different university, agreed. "I think people in general are smart enough to know how to manipulate the volume of a personal stereo," she said. "But a rifle shot, a gunshot—those things are at 140, 150 dB. Your ear doesn't have a chance." Hunters sometimes say that they can't wear ear protection because they need to be able to hear things like deer walking through dry leaves—although, of course, hunters who have deafened themselves by hunting can't hear those things, either. Corey told me that when he and his wife were interviewing for jobs at Massachusetts General Hospital, many years ago, the head of the Neurology Department commented that he and his father both had severe hearing loss, but only on the left side. Corey wondered how it was possible to have a hereditary hearing loss that was unilateral, and worried that that he might fail the interview. "But then the neurologist kindly revealed that he and his father were both hunters, and so their left ears were closer to the barrel," Corey told me. (Corey's interviewer later became the dean of Harvard Medical School; they remain close friends.)

Using decibels to categorize noise dangers can be misleading, at least to nonscientists, because the numbers can make potentially catastrophic increases seem like no big deal. The difference in sound intensity between a quiet room in your house and the interior of a moving car is roughly 30 dB—which is about the same as the difference between a front-row seat at the New York Philharmonic and a front row seat at a Megadeth concert. But the difference in the first case is one you might barely notice consciously, while the difference in the second could be the difference between enjoying an uplifting evening and damaging your ears, especially if Megadeth plays encores.

THE THREAT TO HEARING posed by sound increased when our prehistoric ancestors invented tools. Later, and for centuries, the human activity that was the most likely to cause deafness was probably pounding on metal. Blacksmiths, armorers, and ringers of cathedral bells were among the first serious occupational sufferers of what William Cullen—an eighteenth-century Scottish medical-school professor, the personal physician of the philosopher David Hume, and the inventor of artificial refrigeration—called *dysecoea*, from Greek words meaning something like "bad hearing." In 1802, Andrew Ferguson, another Scottish physician, wrote a letter to the *London Medical and Physical Journal* in which he reported that he had encountered four cases of hearing loss among blacksmiths—a cluster that had surprised him, he wrote, although you would think that by then the connection between noise and deafness would have been incontrovertible. Still, Ferguson was an excellent observer. The blacksmiths' loss had "creeped on them gradually," he wrote, and had manifested itself initially as an insensibility to "weak impressions of sound." At first, the blacksmiths "were not so perceptible themselves of this dullness, as those with whom they held conversation." In time, though, all four noticed not only increasing deafness but also "a ringing and noise in their ears"—tinnitus—sometimes accompanied by vertigo and headache.

Ferguson cleaned and examined the ear canals of all four blacksmiths but found no obstructions or other obvious evidence of disease. "I therefore considered the deafness to be owing to a paralytic state of the auditory nerves, occasioned by the noise of forging, to which all of them had been in the daily practice of, for many years." The treatments he selected for one of the cases—"electricity, blistering, spiritous applications to the cranium, and the injecting into the ears oxygenous air"—are less modern-sounding than his diagnosis, but he did observe that the hearing of the treated patient improved when he stopped working as a blacksmith. (The other three patients had no choice but to continue working, "and therefore little was attempted for their recovery.")

The Industrial Revolution, which was driven by the proliferation of very large, very loud machines, was devastating to ears. An exceptionally harmful occupation was building the boilers that powered the steam engines that made everything else possible. In fabricating a boiler, some workers, often young boys, had to crouch for hours inside what was in effect an iron amplification chamber; their job was to secure rivets being pounded with hammers on the other side, and as they did that they endured a noise that would often have been near or above the threshold of pain. In 1886, Thomas Barr, a British physician, compared the hearing of a hundred shipyard boilermakers, a hundred shipyard iron-molders, and a hundred shipyard letter carriers—plus a control group of a hundred men who didn't work in a shipyard and were therefore presumed to have undamaged ears. He found that the controls could hear a ticking watch if he held it six feet away, but that the boilermakers couldn't hear it until he had moved it to about six inches. He also found that seventy-five of the hundred boilermakers were either entirely or substantially unable to hear what was being said at a public meeting. (And these, for the most part, were men who worked on the *outside* of boilers.)

Barr's findings, and those of similar observers, didn't necessarily change the way anyone did anything. In 1902, Sir Thomas Oliver, who was a pioneer in the study of occupational health hazards, concluded, "There is nothing I know of that will prevent boilermakers' deafness, short of substitution of machine for hand riveting." Ah, well. At the time—and for decades to come—employers' concerns about job-related hearing loss had less to do with their workers' well-being than with their continued productivity. In an article in the *Journal of the American Medical Association* in 1942—more than fifty years after Barr's boilermaker study—C. C. Bunch, a professor in Northwestern University's School of Speech, wrote, "Recently an employer was asked what his reaction would be if someone asked permission to test the hearing of his employees. He replied that he preferred to let sleeping dogs lie." And even when workers themselves had some understanding of how to protect their hearing (cotton wool, fingers in ears), they often remained careless.

The history of capitalism is full of examples of workers effectively glorifying the neglect of the people who employ them. In the 1970s, a friend of mine made a film about a furniture factory in Hickory, North Carolina. The workspace was a filmmaker's dream: old tools and veteran craftsmen and ancient workbenches and sunlight slanting through high windows onto expanses of mellowed wood. But my friend had to be careful with close-ups because virtually all of the woodworkers were missing fingers. "It was something of a badge of honor," he told me. "We heard a number of these guys claim that they'd finished the shift in which they cut off the finger. Only weenies still had all ten, and that mindset had been handed down through at least a couple of generations." Many of those same workers had undoubtedly become hard of hearing as well, but to have worn ear protection in such an environment would have been unthinkable. Hearing loss, in many occupations, has often been treated by its sufferers not simply as an unavoidable hazard but as a point of pride.

A DEVELOPMENT THAT WAS EVENTUALLY WORSE for ears than the Industrial Revolution was the invention of gunpowder in China more than a thousand years ago—and the hazards to hearing increased as gunpowder's uses evolved from fireworks to weaponry, and then again as the weapons became more powerful. I once took a tour, in Boston Harbor, of the USS *Constitution*, better known as *Old Ironsides*, and marveled at the trim, spacious gun deck, with its twin rows of shiny black twenty-four-pounders, fifteen on each side of the ship. But I never thought, then, about what that low-ceilinged space must have sounded like during an actual broadside, in a naval engagement. The battle scenes in *Master and Commander*, the 2003 film based loosely on Patrick O'Brian's Aubrey–Maturin novels, give an unsettling sense of how loud and confusing naval warfare must have been. The Age of Sail, for seamen, was also the Age of Deafness.

Yet there are only isolated accounts of hearing loss suffered by nineteenth-century naval combatants—presumably because to complain about invisible wounds would have been unmanly, if not treasonous. Besides, no one cared very much about the health of ordinary seamen, as long as they were still ambulatory. But problems had to be close to universal. The heaviest artillery exchanges often took place between ships separated by yards, if not feet, and at sustained sound levels sufficient to cause pain and permanent damage. "The havock produced by a continuation of this mutual assault may be readily conjectured by the reader's imagination," Louis de Tousard wrote, in 1809, in *The American Artillerist's Companion, or, Elements of Artillery*, a three-volume textbook. "Battering, penetrating and splintering the sides and decks; shattering and dismounting the cannon; mangling and destroying the rigging; cutting asunder or carrying away the masts; piercing and tearing the sails so as to render them useless; and wounding, or killing the ship's company." Tousard was French. He served under Lafayette during the American Revolution, and was imprisoned briefly, in France, during the French Revolution. He later returned to military service in the United States, and in a paper he submitted to the U.S. secretary of war, in 1798, he proposed the creation of a national academy for military officers. West Point— where *The American Artillerist's Companion* was required reading—was founded four years later, substantially on his plan.

The first large-scale uncontrolled experiment in the effect of gunpowder on human hearing was the American Civil War. Many military historians have observed that the carnage in that conflict was as great as it was because the technology of warfare had outrun conventional thinking about tactics; the impact of those same technological changes on soldiers' ears is far less commonly discussed. Civil War reenactors nowadays almost always wear ear protection—high-quality earplugs and protective earmuffs are visible in just about every photograph of shooters of all kinds on the website of the North-South Skirmish Association—but actual soldiers did not. A Confederate colonel who had exhibited what appeared to be uncommon bravery during fighting near Charleston in 1863, by riding without concern through a barrage of artillery shells, explained later, "I didn't dodge, sir, because I am so deaf that I didn't hear them before their explosion."

The nearly universal casualness about protecting the hearing of soldiers was reflected in widely held attitudes about those who had been deafened in battle. In 1887, the General Assembly of Georgia passed an act establishing annual benefit payments to Georgia residents who had been permanently injured during military service in the Civil War. Veterans who had been blinded were entitled to $150 a year, while those who had lost all hearing got just $30—the Georgia Assembly's answer to the deaf-or-blind conundrum. And the compensation for total deafness was the same as that for the loss of one eye. (Veterans who had lost the use of one ear got nothing.) A retrospective study of Union Army veterans who applied for pensions after the war found that a third of them had received compensation for hearing loss, and that in 70 percent of those the main loss was on the left side—as would be expected of right-handed riflemen. (This seems counterintuitive to most people, but most of the sound made by a rifle comes from its muzzle, to which the left ear of a right-handed shooter is usually more exposed. A shooter's right ear is turned away slightly and also protected by "head shadow," the sound-absorbing property of the skull and its contents. Violinists and violists also usually lose hearing on the left side first, and for the same reasons.) Missing from the Union Army study, of course, were soldiers who hadn't lived long enough to apply for pensions.

In 1907, in England, Arthur Cheatle, a military ear surgeon, gave a lecture to a meeting of naval officers, "Gun Deafness and Its Prevention." He said, "In King Edward VII's Hospital for Officers, Sister Agnes tells me that the majority of the naval officers passing through her hands are more or less deaf." He said that a former director-general of the Naval Medical Service had objected that he had heard few "complaints of gun deafness" from officers—but Cheatle observed that someone with the power to remove an officer from active duty would be the last person in whom an officer with a hearing problem would be likely to confide. And, as always, the impact on officers was less severe than the impact on the sailors who manned the guns. "I can imagine the crew of a battle-ship after an engagement being so deaf that orders could not be heard," Cheatle said; "indeed, we read in the account of the capture of the 'Variag,' in the Russo-Japanese War, that the [Russian] crew were absolutely dazed and deaf when taken off."

Cheatle presented several tables based on a hearing study done by a naval surgeon, who had examined fifty seamen, all in their twenties and thirties. The most severely affected were the sight-setters, gun-layers, gunners, and others whose jobs placed them closest to the blasts; least affected were officers and others not directly involved in the firing. Among the surgeon's descriptions of the men who had the most trouble hearing:

> Deafness said to be due to firing of 6-pounder in a hall with corrugated iron roof. Keeps piece of india-rubber between teeth during gunfire.
>
> Six years deaf after firing 3-pdr. in a corrugated shelter in barracks. Right ear bled. Has felt deaf after firing in the ear since.
>
> Very deaf 9 years, came on while standing alongside barbette of 13.5-inch gun when it was fired on H.M.S. "Empress of India." Was almost stone deaf for 2 days after last gunfire.

A gun-layer (a sailor whose job was aiming artillery) told the surgeon that he didn't think firing had made any difference to his hearing, but he was nevertheless unable to hear a whisper from farther than three feet from his right ear and eight feet from his left—among the worst results in the study. Cheatle said that the deafness of Navy gunners was comparable to that of boilermakers, whose working conditions he had once subjected himself to, briefly, out of professional curiosity. His principal recommendation for naval personnel: snug-fitting earplugs made of moldable "clay fibre." He passed around a sample, sold by Thomas Hawkesley, a London purveyor of medical paraphernalia.

The most remarkable fact about Cheatle's lecture is that his listeners, most of whom were military officers, were surprised by any of it. (Two thanked him afterward for his "interesting remarks.") By that point, soldiers and sailors had been deafening themselves in battle for at least a couple of centuries, yet the idea of making an effort to protect their ears struck their superiors as novel. You would think that functioning ears would have been viewed as a military asset, obviously worth preserving. But that's not the case. Undoubtedly, the negligence was at least partly a consequence of the age-old military disdain for "softness" of any kind. And there were many people who believed that exposure to noise made ears stronger—as though ears were muscles and listening to loud noises were a form of exercise, or deafness an illness that a person could gradually develop resistance to. Still, it's hard to understand.

CHEATLE'S OBSERVATIONS NOTWITHSTANDING, few efforts were made to protect the ears of combatants in the First World War, in which noise itself often functioned as a weapon of mass destruction, both physical and psychological. Trench warfare was especially gruesome. "To the squealing of rats and the deafening hum of flies were added the noise of bullets and shells, the bellowing of dying horses, and the screams and moans of dying men, which were so harrowing that risky rescue operations were undertaken, sometimes more to quieten the wounded than to help them," Leo van Bergen, a Dutch medical historian, writes in *Before My Helpless Sight: Suffering, Dying and Military Medicine on the Western Front, 1914–1918*, published in Dutch in 1999 and in an English translation a decade later. "You could begin to hate men who took too long to die."

Worse were the weapons, which were more powerful and therefore louder than any that had been used in combat before that time.

"During a bombardment the noise was loud enough to split the eardrums and it quite commonly caused permanent hearing loss, especially among gunners," van Bergen continues. "The sound of one shell bursting nearby is deafening, let alone thousands. Many men said you did not so much hear the noise as feel it. . . . Sergeant Paul Dubrulle, a Jesuit, described the misery of a barrage at Verdun. He was caught between walls of noise, walls that advanced towards him and slowly knocked him senseless." Even so, one of the *Oxford War Primers*—pocket-size references published by Oxford University Press and used by military medical personnel in the field—characterized soldiers' claims of noise-induced hearing loss as malingering.

As the war went on, the connection between battle noise and permanent hearing loss became obvious to many in the military. But efforts to prevent or reduce the damage were virtually nonexistent, and where they were tried they were mostly ineffective. One problem was the inadequacy of the available materials; before the development of soft, resilient plastics, making decent ear protectors was next to impossible. But a bigger problem, probably, was the soldiers themselves. Plugging or covering the ears while under fire was stigmatized as cowardly, and soldiers who did so risked not being able to hear what was going on around them—one reason that, even today, soldiers are often reluctant to use ear protection at the times when it would do the most good. And of course a soldier with his fingers in his ears is incapable of holding a weapon. As with previous conflicts, the main official focus was retrospective: how (and whether) to compensate veterans whose hearing had been damaged or destroyed by their wartime service.

During the Second World War, the United States issued soft plastic plugs, called V-51R Ear Wardens, to some artillery crews. They were manufactured by a company in Pennsylvania, they came in three sizes, and they were inserted into the external ear canal by means of a tweezer-like applicator. How many soldiers actually used them, and how effective they were for those who did use them, is impossible to say; the Defense Department was still studying them almost twenty years later. (One conclusion of that study was that V-51Rs provided little or no protection to the many soldiers who inserted them incorrectly, which was easy to do.) Flents Ear Stopples—wax-impregnated cotton plugs used by my mother to soften my father's truly impressive snoring, and by my wife and me when we lived in Manhattan and hated being awakened by garbage trucks at three in the morning—were introduced in the late 1920s, by an American who had observed citizens of Paris using something similar to protect themselves from the noise of the city. Flents would have worked better, and would have been harder to use incorrectly, in addition to being cheaper. And, unlike V-51Rs, they're still sold today. But if any soldiers used them, it was only because they had bought their own.

Despite indisputable evidence that war is bad for ears, threats to the hearing of American soldiers remained mostly unaddressed at all levels of the military. A few years ago, I mentioned hearing loss during a talk I was giving about something else, in Colorado, and a member of the audience told me afterward that when he was in Vietnam he and his buddies used to stuff cigarette butts into their ears; other veterans have told me they did the same with spent brass cartridge cases. Cigarette butts probably make decent impromptu earplugs, but soldiers deserve better. James Henry, a research scientist at the U.S.

Department of Veterans Affairs, told me, "The problem is enforcement. Even seating earplugs—getting them to fit properly—can be really difficult. Soldiers often put them in incorrectly, and don't really get a good seal, so the sound gets through in spite of the plugs. And then, as you can imagine, in combat situations they don't want anything in their ears. So it's a real conundrum for the military. Some situations are straightforward. On a shooting range, soldiers should wear both earplugs and earmuffs, and, at least in theory, that's easy to enforce. But in other situations enforcement is much more difficult." One problem is that, in spite of everything military higher-ups have supposedly learned about hearing loss during recent centuries, the same old biases against "softness" still exist. Just skim any subreddit concerned with hearing loss suffered by military personnel: "My platoon commander called us pussies for wearing ear pro, more or less. He insisted that we can't wear them in combat because we won't be able to hear orders. Granted, I didn't often wear mine in country, but still he was a fucking shitbird. Also, fuck tinnitus."

The wars in Iraq and Afghanistan have been especially hard on the hearing of veterans, partly because combat is louder than it used to be, and partly, perhaps, because improved medical protocols have made wounded soldiers more likely to survive deafening blasts that would have killed them in previous conflicts. (A fifth of all hearing aids sold in the United States are purchased by the Department of Veterans Affairs.) In 2016, Stephen Carlson, who served two tours in Afghanistan, wrote about the effect of his military service on his ears: "The dangers to hearing in the military are too many to count. The number of high-capacity engines involved are legion, from jets to tanks to the massive turbines powering ships. The crack of a standard

M16A2 rifle is 152 dB.... Even electrical generators can inflict serious damage over a long period of time." Simply sleeping on an aircraft carrier—one of the loudest work environments on earth—can
cause permanent hearing loss. Training exercises can be dangerous,
too; Carlson wrote that an "urban warfare course," in which he
learned to use explosives to blow down doors, left his ears ringing for
days. But the worst is actual combat. "By the time the roadside bomb
targeted my vehicle in 2009, my hearing had already been damaged,"
he continued. "During my first tour in Afghanistan, one of my friends
once fired his light machine gun during a firefight outside Bermel,
the muzzle only inches from my ear. It's an incident we joke about to
this day, but it left me practically deaf for a week. Sometime later a
close call with a rocket-propelled grenade only added to the problem,
and that was on top of thousands of rounds of machine gun fire."

In 2002, the military began issuing soldiers two-ended Combat
Arms Earplugs, manufactured by a company that was later bought by
3M. (With one end inserted, the plugs were meant to attenuate loud
sounds while preserving "situational awareness"; with the other end
inserted, they were meant to suppress impulse noises, including gunfire.) The plugs were issued until 2015, to soldiers deployed to both
Iraq and Afghanistan, but were not effective. In 2018, 3M paid $9.1
million to resolve a lawsuit by the U.S. government, under the False
Claims Act, which alleged that the company had "knowingly sold"
defective earplugs to the military "without disclosing defects that
hampered the effectiveness of the hearing protection device." According to a press release from the Department of Justice, 3M and a
company it had bought, Aearo Technologies, knew that the earplugs
were "too short for proper insertion" and that they "could loosen

imperceptibly and therefore did not perform well for certain individuals," yet never revealed the problems to the Department of
Defense (which apparently did no testing of its own). The earplugs
themselves therefore contributed directly to the epidemic of hearing
problems suffered by recent veterans, and online reactions from them
were justifiably scathing: "The pricks who knowingly did this should
be blown from a gun, but as a courtesy they can wear their own earplugs." In 2019, a former Army sergeant sued 3M as well, blaming
the defective plugs for his tinnitus. Hundreds of similar lawsuits
followed.

The Army now issues many combat soldiers an electronic headset
called the Tactical Communications and Protective System (TCAPS).
It's capable of muffling or shutting down many loud sounds, including engine noise and gunfire, while also enhancing soldiers' ability to
hear quiet sounds, and, in one configuration, it can function as headphones with existing radio communication systems. One of my neph-
ews, who is in ROTC and the Army Reserves, has used a TCAPS
headset in training exercises and says it works great. Many hunters
use similar protective devices, which amplify ambient noises but
digitally suppress gunshots after a lag measured in fractions of a
second. A downside is that TCAPS, like any electronic device, needs
to be recharged, and frequently—a challenge in the field. And, naturally, TCAPS headsets are more expensive than most roughly
similar devices sold by companies that outfit hunters—a couple of
thousand dollars versus a couple of hundred. But, even at Pentagon
procurement rates, protecting soldiers' hearing in situ is surely more
economical, in the long run, than providing them with medical treatment, hearing aids, and compensation later. (Hearing-related

disability payments to veterans in fiscal year 2010 amounted to almost a billion and a half dollars, according to the Department of Defense.) In 2017, the Marine Corps began using weapons with built-in suppressors, better known as silencers, to protect their hearing and also to make it easier for them to communicate with one another during firefights. Of course, the suppressors affect only their end of any engagement and do nothing to muffle machinery or other weapons or the explosions of improvised explosive devices (IEDs). What's truly remarkable is that this problem, which was first accurately identified by military personnel in the 1700s, has yet to be satisfactorily resolved.

Five

CICADAS IN MY HEAD

James Henry, the VA research scientist I quoted in chapter four, shares an affliction with millions of military veterans: tinnitus, a constant ringing in the ears. "I played ukulele, and I started playing guitar probably in the third grade," he told me. "I got an electric guitar when I was in junior high school, and I played in surf bands." In the late 1960s, he became the lead guitarist for Eli, a Florida-based group that, after he quit, performed as a warm-up act for Kiss. (They're still around, although with mostly non-original personnel.) "I supported my family that way," he said. "We were busy every week, and we did a lot of traveling, so I was exposed to a lot of loud music. I remember going home at night and having this roaring in my ears. I knew it was caused by the noise, but I didn't realize that the roaring would eventually become a permanent condition." He's in his early seventies now, and his tinnitus, which is severe, has never gone away. "It's high-pitched—very high-pitched—and it's always there," he said. "It's like a super-high-frequency tone, and it's very loud. I can hear it in almost any situation, even in a noisy environment."

That's not what led him to his profession, though. Henry and his wife have a daughter who was born with virtually no hearing—because of a genetic problem, not because of Eli concerts experienced in utero. In the early 1980s, they moved to Portland, Oregon, so that she could attend the Tucker Maxon School, which specializes in teaching speech and listening skills to deaf children. That experience prompted Henry to earn a master's degree in audiology, and after he went to work at the VA he got a PhD in behavioral neuroscience. "I'm a full-time researcher at the VA, and I've been doing tinnitus research for about twenty-five years," he said. "Our focus is on clinical management. There's no proven cure for tinnitus, unfortunately. It can sometimes resolve spontaneously, but, usually, if you've had it for six months or longer it's considered a permanent condition." Henry is in exactly the wrong profession, since when he's at work he's almost always thinking about ringing ears and is therefore seldom able to ignore his own. The same thing happens to veterans he treats. He told me, "When we bring in people who have tinnitus as research participants, they sometimes say, 'Well, I wasn't really thinking about it before—but now I am.'"

IN THE FALL OF 2006, I traveled to Beijing on a reporting assignment. I took a taxi from the airport to my hotel, in the center of the rapidly expanding city, and felt like Esther Summerson, in *Bleak House*, as she approached London for the first time: "I was quite persuaded that we were there, when we were ten miles off; and when we really were there, that we should never get there." The air was pretty Dickensian, too. The smog got so thick during the week of my visit that flights were delayed, sections of three expressways were closed, outdoor school activities were canceled, and buildings across the street from my hotel were visible only as silhouettes in a mustard-brown miasma. "This is unusual for November," a resident told me. "Ordinarily, it only gets like this in August." I spent much of my free time exploring the city's thousand-plus-year-old alley neighborhoods— its *hutong*s, which the Chinese were then demolishing to make way for expressways and shoddy-looking Soviet-style apartment blocks. I visited parks wreathed in coal smoke and automobile exhaust, and watched (mostly older) people practicing ballroom dancing and working out on brightly painted outdoor exercise equipment. There was no escaping the pollution. I imagined that I saw yellow fumes fingering past the curtains in my hotel room—or maybe I really did see them. I caught a bad cold, which got worse on the long flight home, and then got much worse. I felt as though someone had poured concrete into my head and was now gradually tightening a belt around my temples. My sinuses didn't fully clear for a month. Eventually, I stopped coughing. And, when I did, I noticed a ringing in my ears.

At first, I assumed that the ringing would go away, as my cold eventually had. But it didn't. After six months of fluctuating anxiety, I made an appointment with my doctor. "Tinnitus," he said. (Medical professionals, almost without exception, place the stress on the first syllable, as they also do with *angina*; civilians, in both cases, tend to stress the second and to make the *i* in the middle long.) Tinnitus is usually accompanied by hearing loss—and people who have both sometimes assume that it's the tinnitus that's making it hard for them to hear. Tinnitus can indeed make concentration difficult, especially in quiet environments, but it's a consequence of hearing loss, not a cause. My internist tested my ears by holding up a vibrating tuning fork and asking me to tell him when I could no longer hear it. After a while, he leaned forward to make sure the tuning fork was still humming, since he himself could no longer hear it. (We're about the same age.)

There are websites that enable you to compare various examples of what ringing ears sound like to people who have them, and by playing three of the sound files simultaneously and adjusting their volumes I was later able to create a facsimile of my own problem. Those three files sounded like: the hum of some high-tension power lines that transect a golf course my friends and I sometimes play; the buzz of a ceiling full of dimmed halogen lamps; and the drone of the cicadas I listened to on sweltering summer nights when I was a kid. My personal intracranial symphony!

Because the sound in my head seemed to me to be mostly or perhaps entirely on one side, my left, my doctor worried that the cause might be an acoustic neuroma, also known as a vestibular schwannoma, a benign tumor that grows on the auditory nerve. To check, he ordered an MRI at our local hospital. My health-insurance policy has a ruinously high deductible, because I'm self-employed and live in the United States, and the hospital at the time was aggressively amortizing a major investment in fancy radiology equipment, so my first thought was "If the tumor is merely *benign* . . ." But it turns out that you don't want even a nonmalignant growth on your auditory nerve, and if you do have one you want to deal with it as quickly as you possibly can, in the hope that doing so will prevent it from deafening you and causing any number of additional problems.

Of course, the surgery is risky, too—as messing around with any part of the auditory system always is. When the comedian and talk-show host Stephen Colbert was a child, doctors removed a different kind of benign tumor, a cholesteatoma, which had virtually swallowed his right eardrum. To get at it, they had to partially detach and fold forward his pinna—the visible outer part of the ear—and, because the operation altered the contours of his head on that side, his reattached right ear looks slightly different from his left. (He can now tuck the top of his pinna into his ear canal and then, by sort of wincing, make it pop back out—a Stupid Human Trick that he once performed on TV for David Letterman.) Colbert's surgery was successful in the sense that doctors were able to remove the tumor—in 2005, Colbert told my *New Yorker* boss, David Remnick, that "they scooped it out with a melon baller"—but the growth was so extensive that the operation left him with no hearing on that side. "Now I can't get my head wet," he said elsewhere. "I mean, I can, but I can't really scuba dive, or anything like that."

My MRI was so expensive that I was perversely disappointed when it showed my brain to be cancer-free—although when I studied the image on the technician's computer screen I was interested to discover that eyes, behind what I now think of as modest slits, look almost as large as tennis balls. Some kinds of hearing loss can be reversed surgically, and when the surgery is successful the associated tinnitus, if there is any, usually goes away, too. But I didn't have that kind of hearing loss. Tinnitus is sometimes caused by earwax, and can be cured by its removal. Charles Liberman told me, "What is almost certainly going on there is that almost everybody can hear some phantom sounds if the environment is quiet enough, and making the outside world quieter by obstructing the ear canal, or the middle-ear bones, can make us more aware of these phantom sounds." But I didn't have an earwax or a middle-ear problem. There's a form of tinnitus that manifests itself as rhythmic pounding, throbbing, or whooshing, typically synchronized with the heartbeat. It's called pulsatile tinnitus, and, unlike all other forms, it's sometimes audible to people who are not its sufferers. (One way doctors check for it is by placing a stethoscope over an affected ear.) Pulsatile tinnitus can often be eliminated with drugs or surgery. But I didn't have pulsatile tinnitus, either—although I've experienced it, on occasion, for brief periods. My main course of action, my doctor said, was to go back to doing what I'd gotten pretty good at doing during the months before I finally got around to seeing him: making my best effort to pay no attention to the illusory sound in my head. He also said that I should be extra assiduous about using ear protection, in the hope of preventing things from becoming worse.

REACTIONS TO TINNITUS, by people who have it, are highly subjective. Robert Dobie, the professor I spoke with at the University of Texas Health Science Center, told me, "If you define *suffering* from tinnitus as more than just *having* it—if you define it in terms of its effect on your daily life, whether that's distraction or sleep trouble or concentration difficulties or emotional difficulties—then the majority of people who have tinnitus are not tinnitus sufferers." People who are sufficiently bothered to consult a physician are often worried less about the tinnitus itself than about what they suspect may be possible causes. Dobie continued, "More often than not, those people have concerns about, you know, am I going deaf, or am I having a stroke, or do I have a brain tumor. And, after a proper examination and workup, we're able to tell most of those people that they have nothing medically significant, in terms of what they were worried about. That, in itself, is all that some people need." People who need more than that are often helped by brief counseling and an explanation of what the treatment options are, including cognitive, behavioral, and other talking therapies, similar to protocols for people who have phobias. Another researcher told me about a man whose tinnitus was so loud and intense "that it was driving him insane." Initially, he rated his misery, on a scale of 1 to 10, as 10; psychotherapy enabled him to reduce that to 6. Nothing about the noise in his head had actually changed; all that was different was his ability to accommodate it. A celebrated professor of otology and laryngology at Harvard Medical School used to suggest to tinnitus patients that they buy shoes one size too small, so that they'd think about their feet instead.

People who suffer profoundly from tinnitus occasionally undergo "ablative surgery" of the ringing ear, thereby deafening themselves on that side, in the hope of silencing it. But this seldom works, on the rare occasions when it's tried. In fact, severing or damaging nerves related to hearing can itself cause tinnitus or make existing tinnitus worse. When my friend David Howorth first told me that he could hear nothing with his left ear, I said that, at least, by plugging his functioning ear he could experience total silence—but, no, his dead ear rings constantly, and has since the moment it stopped working.

A writer friend of mine has had tinnitus for longer than I have, and over the years he'd gotten pretty good at ignoring it. One day, though, his problem became dramatically more annoying. "I tolerated the hissing, but when the beeping started I went running around the house, trying to find the appliance that was making the noise," he told me in an email. He panicked and made an appointment with a local ear doctor, who referred him to a tinnitus expert in New York City. "This doctor's office was on the Upper East Side, and the walls were decorated with *New York* magazine 'Best Doctors' covers," he continued. "Nice man. Quiet. He listened patiently as I told him about my condition and described everything I'd heard or read about tinnitus. When I finished, he smiled and said, 'Here's the truth. I don't know. They don't know. Nobody knows. What are you doing for it?' I said that when I was unable to sleep I took an Ativan. He said, 'That's fine. You can do that for the rest of your life.'"

John Wawrzonek, a retired electrical engineer, now in his seventies, who, when he was in his thirties, began a second successful career as a landscape photographer, had a truly alarming form of tinnitus. "I used to hear a jet engine, constantly, on one side," he told me. "I've also heard what sounds like groups of people talking. It would come and go, up and down, and I swear that I could almost understand what the people were saying. It was always groups, male groups, and every once in a while I was pretty sure that I heard specific words. It was terrible." (An audiologist told me: "The most common forms of tinnitus are the ringing and the buzzing, but I've also had patients who've said they hear, like, the National Anthem playing in their ear. We tend to call something like that an auditory hallucination, but, really, it's tinnitus, too." A reader told me about his own unusual tinnitus: a "sympathetic" rattling that occurs in response to certain low-frequency sounds, such as those made by air conditioners or passing trucks.) Wawrzonek is also unusual in that, for unknown reasons and against all odds, in recent years his tinnitus has virtually disappeared. "I haven't even thought about it for quite some time now," he told me. "I get a motorboating sound, which popped up for a short while a couple of weeks ago. I'm listening to my right ear now, to see what it's up to. It's gone, and for that I'm very grateful."

So there's at least an atom of hope that tinnitus, after many years, can disappear or fade to insignificance all by itself. But Wawrzonek is one of a vanishingly small group of people I know of who have had that experience.

DURING THE PAST FEW YEARS, I've exchanged many emails with James Gold, a New York City investment banker and consultant a few years older than I am. Our shared interests include golf and hearing problems, although he's probably played more great golf courses than I have, and his hearing issues are orders of magnitude worse. Not long ago, we met in person for the first time, over lunch roughly halfway between my house, in Connecticut, and his weekend house, in Westchester.

We purposely met in a restaurant: the universal venue of choice for experiencing hearing problems, though not necessarily for discussing them. He'd already been seated when I arrived. "We're sitting against this wall," he explained, "because I get to all restaurants ten minutes early so that I can find a wall or choose the table." The room was almost comically ear-antagonistic: loud music, loud customers, crying baby, miserable acoustics. "Being here is like being in Madison Square Garden for me," he said. I had trouble with the noise, too—and so did my digital voice recorder, I discovered later, when I listened to our conversation again. The cacophony enabled Gold to notice something about me that I myself had never noticed: when I had trouble hearing what he was saying over the hubbub in the background, as I often did, I turned my right ear—which I guess I ought to refer to from now on as my "good ear"—toward him.

Gold has had a severe hearing problem since birth. He was given his first standard hearing test, which measures the ability to detect isolated tones along a range of frequencies, when he was five, and it showed a 95 percent deficit in his left ear. That handicap had never caused him serious difficulties, he said, since he'd never known anything else. "Fortunately, I'm left-handed, so it was easy to sit at the end of the table with everyone on my right."

But then his hearing problems became much worse. One day in 1991, his good ear began to waver. "It would be normal for a while, then fade, then return," he said. "So I went to see an ear doctor at New York Hospital." The doctor gave him a standard hearing test, which showed nothing new: his hearing in his right ear was well within the normal range across all frequencies. "The doctor said, 'Maybe you're getting a head cold,'" Gold recalled. "He said, 'Go buy some Afrin, and make sure your nasal passages are clear.'"

The next morning Gold had what he described to me as a complete "vestibular breakdown": his hearing began fluctuating wildly, and he had trouble standing up, and even his eyesight seemed to be off. He called a different doctor, who told him to come to his office immediately. "By the time I got there, I could no longer walk," he said. "I crawled through the door and threw up in the waiting room. They stuck a needle in my arm."

The diagnosis was sudden sensorineural hearing loss, also known as sudden deafness. It's occasionally caused by things like tumors, multiple sclerosis, and strokes, but the vast majority of cases are "idiopathic"—the technical medical term for "Who the hell knows?" Sudden deafness almost always occurs in one ear only, and once it truly begins it almost always becomes full-blown within two or three days. The standard treatment is corticosteroids, which are anti-inflammatory drugs. Gold's new doctor prescribed high doses of oral prednisone. (Nowadays, the drugs are often injected directly into the middle ear.)

"When I got home, I just crashed," Gold continued. "And when I woke up, about three hours later, I was completely deaf—zero response—and the ocean was crashing in my ear. It was a life-changing experience. In a moment, I had gone from being one person to being another person—a person who was deaf." He made an appointment with another ear doctor, who doubled his prednisone dose.

"The prednisone affected me like speed," he said. "I'd sleep just thirty minutes a night, and I was eating a lot, and I basically went crazy." But his hearing began to creep back. With components he bought at Radio Shack he improvised a temporary hearing aid: headphones and an amplifier, with a microphone that he asked people to speak into. By July—eight months after the initial crisis—roughly a third of his hearing had returned, all in the lower frequencies. And that's where his recovery stopped.

It's probably a good thing that Gold's sudden hearing loss was diagnosed early. When steroid treatment is begun within four weeks of onset, patients have an 80 percent chance of recovering at least some of their hearing. No one fully understands why steroids work—or, surprisingly, even whether they do. The drugs have been referred to by some otologists as "holy water." But no one has proven that they don't work, and if I ever suddenly lose hearing I'll certainly demand that my own doctor prescribe them to me.

Because sudden hearing loss almost never affects both ears, people sometimes don't notice it right away—maybe because they have a cold, or because they usually hold their phone to the other ear. And diagnosis is complicated by the fact that several transitory conditions present in the same way. People whose ear canal has become impacted with earwax have similar symptoms: difficulty hearing and the unpleasant feeling that someone has blown up their middle ear like a balloon. An abrupt change in cabin pressure during a descent into Los Angeles International Airport thirty years ago made my right temple feel as though someone had driven an eight-penny nail into it, and it left me virtually deaf and in pain all night and for most of the next day—symptoms that differed from those of sensorineural sudden hearing loss because the (temporary) deafness was bilateral and I was certain I knew the cause.

As is often the case, Gold's sudden hearing loss was accompanied by severe tinnitus, which began the same day and has never gone away. "It's so loud that it sometimes awakens me at night," he told me. "And in the morning I wake up in a sea of noise. There are days when it's so oppressive that I basically fake my way through. More often than not, the sound is sort of the hiss of a broken fluorescent lamp, at a frequency between about twenty-five hundred and thirty-two hundred hertz. There's also a burning sensation—actual pain. Every once in a while, the sound will go from the hiss to a pure tone—a piercing, very narrow noise—but then it breaks up. That happens maybe three times a year, for a minute. And that's fortunate, because the people with pure-tone tinnitus are the ones who end up killing themselves."

THERE ARE CERTAIN FREQUENCY RANGES that Gold often hears at multiples of their normal volume—an auditory problem called hyper-acusis. Severe cases, in which the increased volume is accompa-nied by physical pain, can be more devastating than deafness. I met one such sufferer in early 2019. When I arrived at his house, in a suburb of a large eastern city, I worried about ringing the doorbell. Then I noticed two rectangles of dried, blackened adhesive on the doorframe, just above and just below the button. I deduced that the button had been taped over at some point but was now safe to use. I pressed as gently as I could, and, when the door opened, was greeted by a couple in their sixties and their son. The son has asked me to identify him only as Mark, his middle name. He's thirty years old, and he's tall and trim, and on the day I visited he was wearing a ma-roon plaid shirt, a blue baseball cap, and the kind of noise-deadening earmuffs you might wear while firing an assault rifle at a shooting range.

Mark and I sat in the living room, at opposite ends of a long coffee table, and his parents sat on the couch. He took his earmuffs off, but he didn't put them away. "I was living in California and working in a restaurant," he said. "Somebody would drop a plate or do something loud, and I would have a flash of ear pain. I would just kind of think to myself, Wow, that hurt—why was nobody else bothered by that?" Then, suddenly, everything got much worse. Quiet sounds seemed loud to him, and loud sounds were unendurable. Discomfort from a single exposure could last for days. He quit his job and moved back in with his parents, on the East Coast, after a cross-country flight during which he leaned all the way forward in his seat and covered his ears with his hands.

That was five years ago. Hyperacusis can be caused by overexposure to loud sounds, although its exact etiology remains mysterious and no one knows why some people are more susceptible than others. (Lyme disease is one of a number of possible contributing factors.) As with tinnitus, which Mark also has, there is no cure. Before the onset of his symptoms, Mark's life was noise-filled but not significantly different from the lives of millions of his contemporaries: garage band, earbuds, crowded bars, concerts. The pain feels like "raw inflammation" and is accompanied by heavy pressure on his ears and temples, and by tension in the back of his head. His tinnitus is similar to the type suffered by John Wawrzonek. "You and I would have a conversation, and then after you'd left I'd go upstairs and some phrase you had been saying would repeat over and over in my ear, almost like a song when they have the reverb going," he said. "That doesn't happen all that much now, although it still does every once in a while."

Mark now manages his symptoms better than he did five years ago, and he sometimes does simple errands in his parents' car, but his life remains circumscribed. He hasn't worked since he left California. The day before my visit, he had winced when his father crumpled a plastic cookie package before dropping it in the recycling bin. Toward the end of our conversation, which lasted an hour and a half, he put his earmuffs back on. His head, he told me later, still hurt that night.

I HAVE WHAT'S USUALLY DESCRIBED as a personality type that's useful in dealing successfully with tinnitus: when I learned, from my doctor and from Google, that there was nothing I could do to make it go away, I thought: Good—then I'll do nothing. And when I began researching tinnitus in order to write about it, and therefore couldn't avoid thinking about it and occasionally mentioning to my wife something I'd learned, over dinner, she marveled that, during the previous decade, I'd never complained about it to her. The main source of my stoicism is probably just laziness. Still, my laziness has been therapeutically useful. The hopelessness of tinnitus is a bummer in some ways, but in other ways it's a relief. If I could cure my tinnitus by losing thirty pounds (let's say), I wouldn't necessarily be happier, because then I would be angry at myself not only for having damaged my ears but also for failing to lose the weight. When I described this (healthy, reasonable, emotionally balanced) attitude to Gold, however, he was appalled. "Bad answer," he said.

Gold's approach to his own tinnitus has been dramatically different. Partly that's because his symptoms are worse, but it's also because he's the sort of person who is constitutionally incapable of taking no for an answer—the same personality profile that has made him successful in his profession. He also has the financial resources to pursue just about anything he feels inclined to pursue. And he has pursued his hearing problems.

"When you have chronic issues that you're not likely to die from, you have to be your own advocate," he told me. "With tinnitus, I started from a point at which I could not actually comprehend that the best doctors in the world couldn't restore things to what they had been. It took me a couple of years to fully understand that, no matter where I went and no matter how much time and money I dedicated, I wasn't going to wake up one morning and find out that this whole matter had been a bad dream."

He didn't give up, however, and there have been extended periods during which his tinnitus was the focus of his life. He continued, "I've followed a countless number of dead ends, including three weeks in Jerusalem working with some whizbang guy who was said to be able to 'cure' the tinnitus of Israeli pilots and soldiers who'd blown out their hearing on the battlefield. But that proved to be nothing, at least to me." He traveled to Beijing in order to be examined at an alternative-medicine center he'd learned about. "It was a huge place, a dozen or more stories tall, and globally renowned for its use of acupuncture to treat ailments that Western medical practice can't," he said. "I underwent several hours of examinations and testing and interviews, and after all that I met with a group of very wise Chinese doctors, the youngest of whom looked about ninety. I'm sure that much of what they told me was lost in translation, but the essence of it was that, although they could deal with virtually everything else, even after thousands of years they didn't have a clue about tinnitus." He sought treatment in Atlanta, Baltimore, London, Minneapolis, and Los Angeles. He spoke by telephone with experts in Oregon, Australia, and Argentina. "Back in the days before you could search for things on the Internet, I was getting faxes from the American Tinnitus Association," he said. "The president of the association came to see me, and he said that they wanted me to go on the board, and at one point they offered to make me the association's president."

In the mid-1990s, Gold traveled to Memphis, to be treated at the Shea Ear Clinic, a practice established nearly a century ago by John Shea Sr., who died in 1952. At the clinic—which was then being run by Shea's son, John Shea Jr.—Gold underwent a tinnitus treatment that had been developed there, in which injections containing lidocaine, a drug that dentists use as a local anesthetic, temporarily reduce or eliminate tinnitus, apparently by (in effect) numbing the malfunctioning parts of the auditory system. "John Shea had written an article about injecting lidocaine directly into the ears of tinnitus patients," Gold told me. "I didn't let him do that, but his associates introduced me to the intravenous use of lidocaine." The drug was mixed with saline solution and administered as a slow drip in several sessions over three days. "In my case, the lidocaine eliminated my tinnitus for the ninety minutes I was on the drip," Gold continued. "It was the first time in four years that I had experienced true silence. The noise would just go away. Gone. Wow. Unbelievable."

Lidocaine infusions aren't effective for all tinnitus sufferers, and among those for whom they are effective the relief isn't permanent. "And you can't walk around with a lidocaine drip," Gold continued. At that time, there was no such thing as a lidocaine patch, because of the difficulty of making the drug move through the skin. (Recent research suggests that won't always be true.) "My experience at Shea led me to a number of lidocaine analogs, which I would try serially," he continued. "Some of them would work a little bit, and some of them would work not at all."

A recommendation commonly made to people with tinnitus is to stop drinking coffee.

"Are you on caffeine?" Gold asked me.

I said I was.

"Get off," he said. "Most people will say it doesn't make a difference, and there's actually no evidence that cutting it out helps. But the idea is to get rid of all the stimulants."

I told Gold that, if I were forced to choose between coffee and total tinnitus relief, I might choose coffee.

"Ah, you're a sick puppy," he said.

IF YOU SPEND ALMOST ANY TIME reading about tinnitus online, your email in-box soon fills with spam about the surprising food that, if you eat it for breakfast, will make your ears stop ringing forever, or the homeopathic concoction whose main selling point is that audiologists and the FDA don't want you to know about it, or the treatment that "leaves doctors speechless," or the one that makes brain surgeons "scream this is 'medically impossible,'" or the one, from "Dr. Tinnitus, M.D.," that will cure tinnitus in less than seven days. The first time I received an email with the subject line "Reverse Tinnitus," I thought (well, hoped) that it would describe a condition that causes people's ears to make an annoying noise which is audible only to other people. (That's not what it meant.)

Reading and thinking about tinnitus has also made me hyperaware of the nature of mine. Every so often, the volume and pitch will increase noticeably, usually in just one ear, and sometimes the sound suddenly seems louder than I remember its having been before. So far, those changes have been transitory; are they real, and therefore something to worry about, or are they just products of excessive contemplation? One autumn afternoon, the noise in my ears not only got much louder but also began to oscillate, and I panicked. A little later, though, I noticed that the new sound changed when I turned my head and that it became faint when I covered my ears—the opposite of what happens with tinnitus. Then, when I moved to a different room, the noise went away entirely, and I realized that what I'd been hearing wasn't malfunctioning circuitry in my brain but an insect of some kind, either inside my house or just outside a window. Whew! When a *New Yorker* editor of mine was reading something I'd written about tinnitus, he suddenly noticed, with alarm, a ringing in his own ears. Eventually he realized that the sound was coming from an adjacent office, where another editor, who had also read what I'd written, was listening to online tinnitus simulations. An audiologist I spoke with told me that she and her colleagues had recently had a discussion about whether they really ought to be asking tinnitus patients to keep tinnitus diaries between appointments. "In one way, it helps you understand what they're going through," she said. "But it also makes them focus on their problem a lot more, and it's not clear that that's really going to make it easier for them to manage it."

Treatment for tinnitus often includes hearing aids, which can disguise the problem, in people who've lost hearing, by bringing up the volume of everything else: a researcher I spoke with likened tinnitus to a candle in a darkened room, and said that one way to make the candle less noticeable is to turn on some lights. Another way to make tinnitus less bothersome is to mask it with real sound. Many people with tinnitus sleep with fans, air conditioners, or white-noise machines; others use devices or apps that play sounds specifically designed to serve as maskers. The similarity between my tinnitus and the high-pitched droning of insects is seasonally useful. When crickets and cicadas are at their loudest, during late summer and early fall, I often don't notice my tinnitus at all. Somewhat similarly, I occasionally pretend that the ringing in my ears is a sound I play on purpose to mask the ringing in my ears. In other words, instead of playing a masking sound that, if it were the sound made by my tinnitus, would annoy me as much as my tinnitus does now, I just pretend that my tinnitus is a sound I'm playing to cover up my tinnitus—a Zen-like switcheroo that doesn't always work but that, when it does work, saves me from having to get out of bed to turn on a fan. I also once discovered, while walking our dog a few blocks from my wife's parents' house, an entire street on which the volume and pitch of the road noise from a nearby highway exactly canceled my tinnitus. So if I ever got desperate I could suggest that we move there.

My sister sent me a link to an online demonstration of a technique that involves placing the palms of your hands over your ears and repeatedly snapping your index fingers against the back of your neck just below the base of your skull. (Search YouTube for "Reddit Tinnitus Cure.") For some people, the snapping produces a few minutes of silence, although it did nothing for me. Something I've noticed about myself is that, although I can hear my tinnitus all day long if I focus my attention on it, even in noisy environments, to the best of my knowledge I've never heard it in a dream: my ears don't ring when I'm running around in my subconscious missing flights, being chased by bad guys, losing my wallet or my children, or trying to remember the names of college courses I've signed up for but forgotten to attend. I've asked other people who have tinnitus about this, and so far none of them, including James Gold, has remembered hearing their ears ringing in a dream. Does that mean the brain knows how to completely silence itself? If it does, perhaps we could persuade it to share the secret with us when we're awake.

GOLD TOLD ME THAT, although lidocaine hadn't proven to be a cure, its temporary effectiveness had given him a likely insight into the true nature of tinnitus. "It's a chronic pain," he said. "And when I realized that I started talking to the pain guys, who showed great interest, including a guy at Memorial Sloan Kettering, in New York, who was willing to give me high doses of lidocaine once or twice a week. The only problem was that it didn't last."

Current thinking about tinnitus suggests that Gold is correct about the pain connection. An increasingly common theory is that tinnitus is analogous to phantom limb pain, the sometimes intense discomfort that many amputees perceive in parts of their body that are no longer there. Robert Dobie, the Texas professor I spoke with, told me, "You know, a guy loses a leg and still has an itchy toe even though the toe is gone. I think of tinnitus the same way. You've got some hearing loss, and your brain is no longer getting sound input in certain frequency regions, so the brain replaces the silence with a phantom sound."

Phantom limb pain includes not just pain but every sensation that people with intact extremities experience, including heat, cold, wetness, itchiness, soreness, movement—everything. Vilayanur S. Ramachandran, a neuroscientist on the faculty of the University of

California, San Diego, devised a brilliantly creative treatment more than a decade ago. My *New Yorker* colleague John Colapinto wrote about Ramachandran in 2009, and described his interaction with a patient who had lost most of his left arm and now suffered from the perception that the arm was not only still there but also painfully cramped:

> Ramachandran positioned a twenty-inch-by-twenty-inch mirror upright, and perpendicular to the man's body, and told him to place his intact right arm on one side of the mirror and his stump on the other. He told the man to arrange the mirror so that the reflection created the illusion that his intact arm was the continuation of the amputated one. Then Ramachandran asked the man to move his right and left arms simultaneously, in synchronous motion—like a conductor—while keeping his eyes on the reflection of his intact arm. "Oh, my God!" the man began to shout. "Oh, my God, Doctor, this is unbelievable." For the first time in ten years, the patient could feel his phantom limb "moving" and the cramping pain was instantly relieved.

The patient repeated the exercise for ten minutes a day for a month, and as he did his bothersome phantom limb steadily shrank—a tremendous relief, and a transformation that Ramachandran characterized as a virtual "amputation."

Atul Gawande—another *New Yorker* colleague, and a physician—has described phantom limb pain as the brain's "best guess" as to what's currently going on with a body part that it used to be in constant communication with but can no longer detect. In Gawande's interpretation, Ramachandran's mirror gave the amputee's brain a different take on the condition of his missing arm, leading the brain to modify the perceived sensation. "The brain has to incorporate the new information into its sensory map of what's happening," Gawande wrote. "Therefore, it guesses again, and the pain goes away." In effect, Ramachandran's mirror fooled his patient's brain into fooling itself in a different way.

If the auditory centers in a tinnitus sufferer's brain are also making their best guess about the nature of nerve signals they're no longer receiving, could they be fooled into thinking they hear something else—perhaps silence? Desyncra, a German medical company, offers a "neuromodulation" tinnitus therapy, which, it says, "rewires the brain" by altering the function of misbehaving brain cells so that "the brain unlearns to produce the pathological behavior and sustained effects occur." You visit a participating audiologist, who uses proprietary software to pinpoint the frequency of your tinnitus. Then, several hours a day for several months, you wear hearing-aid-like earphones connected by wire to an iPod Touch (they still exist!). The iPod Touch plays sequences of four tones, two just above and two just below the tinnitus frequency, at a volume that's audible to you but not so loud as to interfere with your normal activities. According to Desyncra, people who follow the protocol report significant, lasting relief.

I learned from the company's website that the participating audiologist closest to where I live is a hundred miles away. But even if he or she were right next door, I wouldn't rush over. The cost is significant—$4,500—and it's hard for me to believe that there's anything about either the technique or the hardware that needs to be that expensive. You can estimate the frequency of your tinnitus yourself online, and you can do that every bit as accurately as an audiologist can, since only you can perceive the sound in your head. When Desyncra applied to the FDA for premarket approval, the FDA ruled that it had done so unnecessarily because "the device is substantially equivalent . . . to legally marketed predicate devices marketed in interstate commerce prior to May 28, 1976, the enactment date of the Medical Device Amendments, or to devices that have been reclassified in accordance with the provisions of the Federal Food, Drug and Cosmetic Act." My guess is that, if the technique really does work, you'll soon be able to download and self-calibrate an app that does the same thing for much, much less. If you're feeling adventurous you can try it for free, right now, using a Web app that a musician who calls himself General Fuzz created, based on the description of the Desyncra technique in a scientific paper available on the website of the National Institutes of Health. I'm going to wait.

A FLIGHT THAT I WAS SUPPOSED to be on several years ago was canceled shortly before it should have taken off, for reasons that the airline employees at the gate did not explain. Dozens of stranded passengers crowded in front of the customer service desk to complain. I picked out the most agitated-looking type A businessman and stood as close to him as I could, figuring that, if immediately rebooking a similar flight was possible, he would be the first to find out. In effect, I was outsourcing my vexation to someone who looked like an expert at being vexed. And, when the agent told him (more than once) that she couldn't get him onto another plane right away, I left him alone with his shouting and his throbbing blood pressure, found a quiet seat at an unoccupied gate, called the airline's 800 number, and tranquilly sat on hold.

I now think of Gold as my type A man at the airport. It's useful to know people like this. I have a nerdy friend who tracks the life spans of individual lightbulbs on an Excel spreadsheet, and before I buy almost anything with a microprocessor in it I talk to him, on the assumption that he'll already have done all the necessary product research and comparison shopping. Gold has come to know so many people in the hearing field that, if a real cure for tinnitus ever does emerge, someone he's dealt with in the past will let him know, pronto. And now that Gold and I have spent some time together, I feel the same way about him. And as soon as he's told me, I'll tell you.

Six

CONDUCTIVE HEARING LOSS

My friend Lynn Snowden Picket, who lives in New York City, went to see her doctor because she thought she was coming down with something. In the course of examining her, he looked into her ears. He said that she had some extra wax in her ear canal and that he was going to remove it. Snowden Picket told me, "So he put this tool in my ear—and, *holy mother of God*, was it ever painful! *And loud!*"

Earwax isn't wax; it's cerumen, a sticky substance that consists of secretions from glands that line the external ear canal, plus dead skin cells and odds and ends. It protects the eardrum by trapping and gradually jettisoning insects, dirt, and miscellaneous small intrusions, and also by acting as a mild antimicrobial. Under normal circumstances, it requires no operator maintenance, since chewing, swallowing, and other natural motions cause it to migrate toward open air, where it dries up and flakes away. People with hearing aids do have to pay attention to earwax, because the aids themselves can prevent it from exiting the ear canal; it can also clog the aids' speakers. A friend of mine angrily complained to her audiologist that her new, expensive hearing aids had stopped working, and then she was both surprised and embarrassed when the audiologist restored them to full function by changing their wax filters and cleaning her ears. People who wear earbuds for extended periods sometimes face similar issues.

For almost everyone else, the most common causes of earwax problems are ill-considered attempts to solve or prevent earwax problems. Major culprits are Q-tips, which, when they're used the way most of us use them, function like the long-handled rammers that artillerymen in the olden days used to shove gunpowder and cannonballs down the barrels of cannons. If the compacted plug is large enough, it can cause conductive hearing loss, so-called because it occurs when sound vibrations are prevented from being conducted all the way to the inner ear. The best way to avoid conductive hearing loss caused by earwax accumulations, for most people, is to do nothing.

And yet. Probing your ear canal with one thing or another is hard to resist, especially if—as happened to a relative of mine—your decision to stick an unbent paper clip into your ear was once rewarded by the retrieval of a wax plug the size and shape of a pencil eraser. Visible earwax seems like proof of poor hygiene, not a sign of otic vitality, and the grooming instinct in primates is strong: all those photographs in *National Geographic* of chimpanzees meticulously removing foreign matter from the hair of other chimpanzees. The chimpanzee being groomed clearly enjoys the exercise, but there are pleasures for the groomer, as well. Snowden Picket continued, "I told my doctor to stop, but he said the wax was like a scab, and he was just about finished getting it—could I tough it out?" Once I've begun removing burrs from the fur of my dog, I find it hard to stop, and, because I know the dog will jump from my lap if I pull too hard, I've refined my technique: think like a chimpanzee. The doctor started again. "Now tears were just pouring out of my eyes," Snowden Picket said. "My ear is fucking killing me, and I'm completely spent. What is wrong with people?" (Doctors are primates, too.)

Other blockages are possible. My son went to an emergency room in Washington, DC, late one night after part of a silicone earplug became lodged near the bottom of his ear canal, beyond civilian intervention; the doctor of a friend in California, during a routine ear exam, found a broken-off bit of a forgotten old blue plastic earplug in the friend's ear and removed it; another friend twice lost the dome-shaped ear piece from one of his hearing aids, and had no idea where either one had gotten to, until his doctor extracted both from one of his ear canals. In 2018, a mature cockroach crawled into the ear of a sleeping woman. Fully extracting it took nine days and involved efforts by the woman, her husband, an emergency room doctor, the woman's regular doctor, and an ear, nose, and throat specialist. Similar cases are not unknown. "Roaches are searching for food everywhere," an entomologist told a reporter. "And earwax might be appealing to them." In the relatively uncommon cases in which nature needs help and the intrusion is not a cockroach, the simplest technique is to syringe the ear with warm water or saline solution, then allow it to soak, as you would with a gunked-up baking dish. Cerumen, unlike real wax, is water-soluble.

The worst interventions are the ones that certain kinds of friends inevitably recommend, among them "candling," which involves sticking a wax-impregnated cloth or paper tube into your ear canal and setting it on fire. Candling not only doesn't work—the crud that accumulates in the bottom of the tube has been shown to be mainly residue from the candle itself—but also can burn hair and skin. CVS sells numerous earwax-removal products, and Amazon has pages and pages of potions and paraphernalia, including kits that look as though you could use them to perform dentistry, if not open-heart surgery. There's a toothbrush-size wax-extraction implement for children, listed as a "LuckyStone Kids Baby Safe Ear Wax Pick Ear Cleaner Earpick Spoon Earwax Curette Remover Tool with LED Flash Lighting, Assorted Color." LuckyStone is based in Ningbo, China. Its many dozens of other products include disposable underarm sweat-guard pads, motion-activated in-toilet night lights, and fur-lined handcuffs for sadomasochistic sex play or "girls' night out." Buyer beware.

JEANNETTE BARNES LIVES in Colorado Springs. She's an award-winning poet, and for more than forty years she has also worked as a librarian, mostly with academic, technical, or military collections. When she was a child, in the late 1950s and early 1960s, she often woke up with her ear cemented to her pillowcase—the result, she told me, of "seepage from repeated infections so severe that at last they stopped hurting at all, sealed, nearly dried up, except inside." By the time she was old enough to go to school, she spent two weeks out of four home sick.

Barnes's father spent thirty-three years in the Navy, beginning on a submarine in the Pacific during the Second World War and ending at the Pentagon during the Nixon administration. "If the Navy had wanted its junior officers to be equipped with impedimenta like children, they'd have been issued with a starter kit," she told me. "Military docs in the early sixties were not remotely trained to notice anything about kids." Her tonsils and adenoids were removed, and drainage tubes were inserted into both eardrums, but somehow no one guessed that she was losing her hearing. She would lie next to her dog, a springer-spaniel-dachshund mix, and tell stories or recite poems directly into his ear—the one reliable method of communicating, in her experience. She listened to her parents' radio by pressing her head against its wooden cabinet: Bach, Segovia, Glenn Miller. "I never noticed anything odd about not hearing," she said. "It was just how things were, every day. It must have been quite gradual and, in the end, profound. I could already read, was too sick too often to be around other kids much, and just really did not think anything was unusual."

The idea that someone could be deaf without knowing it seems remarkable to people who don't have hearing problems, but Barnes's case is not extreme. Gerald Shea had scarlet fever when he was six years old, and the infection burned through his hair cells. He lost much of his ability to hear, especially at higher frequencies, yet somehow he made it, with honors, through Andover, Yale, and Columbia Law School. Then he became a partner at Debevoise & Plimpton, a white-shoe New York City law firm, for which he conducted high-level business negotiations in both English and French. He never suspected that there was anything wrong with his ears until, at the age of thirty-three, he was given a routine, job-related hearing test, which showed him to be severely deaf—and when the results were explained to him he was incredulous. "It's no easy task, for anyone, to upset what he considers to be the longstanding, natural patterns of his life," he wrote in *Song Without Words*, a terrific memoir, published in 2013. He'd always assumed that he heard things the way other people did—only somehow not as well, for unknown reasons, and as a consequence was forced to work harder. Without consciously setting out to, he had learned to read speakers' mouths and facial expressions, and to rapidly make educated guesses about consonants he couldn't hear:

> verb science
>> firm science
>>> firm's clients

He became exceptionally adept at lip-reading and extracting content from context, but the effort took a toll. Indeed, the mental exertion required to deduce meaning from diminished signals is one of the reasons that age-related hearing loss can have a devastating impact on the cognitive abilities, social engagement, and general well-being of its sufferers.

When Barnes was eight, her father was transferred to a naval base in San Diego, and before she entered third grade, at a new school, she was given a state-mandated hearing test. A nurse handed her a pair of "itchy, heavy headphones," asked her to sit in a glass booth, facing the other way, and told her to press a button each time she heard a tone. "I sat a long time, poised, thumb on button as I'd been shown," she told me. Eventually, the nurse entered the booth. "Looked straight at me, gestured. Fiddled with knobs. Nothing. I didn't move; no reason to lift my hand. Very slowly, the looks on the faces of the nurse, then the new teacher, changed." As was true with Shea, her deficit had remained hidden for as long as it had because she was uncommonly skillful at outwitting it. "I could read so well that I usually tested way ahead of whatever school-system level we came to," she told me, and she became adept at guessing what people had said. But, also like Shea, she had been cut off from much of what had been going on around her, and had suffered longer than she would have if she'd been less successful at compensation.

Barnes told me that real relief, for her, came not from anything doctors did to her ears but from her father's eventual transfer to a base in the Caribbean. "I think that the beat of light and breezes playing on the ocean baked, at last, all infections out of me," she said. The pain stopped. Her chronic bronchitis went away. Her ears opened up. "It literally was startling how loud ordinary life was, once I could hear it again."

COMPACTED EARWAX AND REPEATED INFECTIONS are common causes of conductive hearing loss, but there are others. When John Wawrzonek—whose severe, motorboat-like tinnitus I described in chapter five—was an undergraduate in the physics department at MIT, in the early 1960s, he noticed that he was having trouble hearing a friend on the telephone. So he moved the receiver from his left ear to his right, and now he heard fine.

He made an appointment at Massachusetts Eye and Ear, which is one of the world's oldest and most distinguished treatment and research centers for hearing disorders. It was founded in 1824 as the Boston Eye Infirmary, a charity clinic whose purpose was "to alleviate suffering of less fortunate brethren." Ears were added in 1827, and an association with Harvard Medical School began in 1866. In 1876, Anne Sullivan—who would later become Helen Keller's teacher—was operated on at the infirmary twice while she was a student at the Perkins School for the Blind, also in Boston. Sullivan had contracted trachoma, a bacterial eye infection, five years earlier, when she was five. The disease had scarred her eyelids and deformed them so that her eyelashes scraped across her corneas, permanently damaging them. The operations temporarily relieved some of her symptoms, but she continued to suffer for the rest of her life.

Wawrzonek told me, "At Mass. Eye and Ear, they gave me some tests, and they said, 'Well, you've got otosclerosis.'" Otosclerosis is a disease of the middle ear. In the type that he had, the stapes—the stirrup-shaped bone that is one of the three auditory ossicles—gradually becomes immobilized, sometimes by bone and sometimes by soft tissue, so that fewer and fewer vibrations make it all the way from the eardrum to the cochlea. Otosclerosis is usually inherited. "My mother had it in both ears, and her mother had it, too," he said. We were sitting in the breakfast room of his house, in eastern Massachusetts. "The earliest possible clue that I had a problem was in grammar school, when we were given hearing tests, and I was one of the few students who got called back. So I surmise that they had noticed something there. But I never knew I had a problem until I was in college." Otosclerosis typically begins in childhood, steadily becomes worse, and eventually stabilizes, usually in early adulthood. "At Mass. Eye and Ear, they told me they didn't want to do anything until the fusing had gone as far as it was going to go." He reached that point in 1970 and underwent surgery on his left ear.

The first reliably successful version of the operation he was given, a stapedectomy, had been performed in 1956 by John J. Shea Jr., a young surgeon at a hearing clinic founded by his father, in Memphis, Tennessee (this was the same clinic, mentioned in chapter five, that later gave lidocaine tinnitus treatments to James Gold). Shea's otosclerosis patient was a fifty-four-year-old woman who could no longer be helped by hearing aids. Shea removed her diseased stapes and replaced it with a minuscule Teflon prosthesis. The surgery was a success. A photograph of Shea at work in his operating room appeared on the cover of *Life* in 1962, in an issue devoted to "The Take-Over Generation." "He never accepts phone calls, hires assistants without formal medical training and during his operations keeps up a steady conversation on anything from brainwashing to the economic insecurity of Baptist preachers," *Life* explained. "He flies his own plane and designs his own surgical equipment. But his satisfactions come mostly from his patients. He was delighted with the one who refused a sleeping pill after an operation because 'she wanted to stay awake to enjoy hearing.'" By the time the story was published, Shea had performed more than four thousand stapedectomies, and claimed a 90 percent success rate. Two years later, he married Lynda Lee Mead, who had been Miss America in 1960. They had five children, one of whom, Paul, first watched his father operate when he was a young boy, and took his grade-school classmates to the clinic on field trips. Today, Paul is a principal in the practice, the Shea Ear Clinic. His father died in 2015, at the age of ninety.

Wawrzonek's operation was performed not by Shea but by another stapedectomy pioneer, Harold Schuknecht, at Mass. Eye and Ear. "You're conscious," Wawrzonek recalled. "They sedate you, and wheel you in. There's a little pain, and you can feel it, but they don't want to put you under because there is a danger of damaging nerves in your face, and they want to be able to see your reaction as they're poking around. You lie there for, I don't know, an hour. It's no big deal." The procedure that Schuknecht performed was actually a stapedotomy, the stapedectomy's streamlined successor, although the older term is still widely used, including by surgeons who perform it.

David Jung is an assistant professor of otolaryngology at Harvard Medical School, and a clinician, surgeon, and researcher at Mass. Eye and Ear. I visited him in his office recently. To me, he looked barely old enough to have graduated from college, but somehow he had managed not only to do that but also to earn both an MD and a PhD (in genetics). He treats patients with the full range of hearing problems, from chronic ear infections to big tumors involving their auditory nerve, and he's also doing research into potential methods of therapeutically reversing sensorineural hearing loss. He performs ear surgery one day a week.

"If I had to name one operation that's the most satisfying operation we do, I'd say it's probably the stapedectomy," Jung said. "You take someone who can't hear, and when the operation is over they can." During the procedure, Jung, aided by a microscope, accesses the middle ear through the external ear canal. He cuts halfway around the perimeter of the eardrum—"like a can opener, almost"—and peels the eardrum back, exposing the auditory ossicles. "We separate the connection between the incus and the stapes, break off the little legs of the arch of the stapes, and use a laser to make a hole in the footplate," he said. The footplate is the flat base of the stapes; it covers the oval window, through which sound vibrations enter the cochlea. When John Shea did the first stapedectomy, he entirely removed the footplate, and replaced it with a graft from a vein in the back of his patient's hand. In the modern version of the operation, the footplate is left in place. "We insert a prosthesis, a tiny piston, into the hole we made in the footplate with the laser," Jung continued. "The piston is typically made of Teflon, and there's a wire that comes off that and hooks onto the incus. Surgeons used to try to re-create the entire stapes, but eventually they realized that they could get the same result with just this little piston, which floats inside the fluids of the inner ear." The piston is roughly the diameter of a standard office staple, if that, and it and its attached wire are not quite as long as one of a staple's legs. The incus moves the piston up and down, like the handle of a well pump.

"In a normal ear, you have the footplate on the oval window, and the arch of the stapes above that, and the only connection to the eardrum is through the incus, on top of the arch," Jung continued. "So the incredible variety of tones and overtones and harmonics that we can hear are all coming from just that little thing going up and down. That has never made sense to me. I sometimes wonder whether there might actually be some kind of three-dimensional vibration of the stapes around its ligament, to account for that. But, however it works, we are able to reproduce it with a zero-point-six-millimeter prosthesis bouncing in and out of this tiny aperture. And the most amazing thing to me is that the quality of sound the patient gets is entirely natural—unlike the tinny, amplified sound of a hearing aid."

IMMEDIATELY AFTER HIS STAPEDOTOMY, Wawrzonek could tell that it had been successful. "My ear was stuffed with cotton and blood clots and so forth, but even through all that things sounded louder," he said. "And then, after they'd pulled everything out, my hearing was normal. The procedure, if it works, restores your hearing, period. And that's what happened with me."

The otosclerosis in his right ear took another decade and a half to stabilize. By the time it had, Schuknecht had retired, and the second procedure was performed by another pioneering surgeon, whom Schuknecht had helped to train. "That ear was down about 50 dB, and I was expecting the same result I'd had with the other," Wawrzonek recalled. But there was no improvement. The surgeon suspected a blood clot but said he couldn't be sure. Two weeks later, Wawrzonek attended a concert at Tanglewood, the summer home of the Boston Symphony Orchestra, on the other side of the state. "The program was Mahler, who wrote some of the most unbelievable symphonies ever—very powerful, very loud. I think it was the Second Symphony that night, and it was coming to a climax, and the entire orchestra was just tearing away. And all of a sudden—wham—my right ear turned on."

That change lasted just two years. Stapes surgery has always carried a small risk of total hearing loss, for reasons that are still somewhat mysterious, and the risk increases with a second procedure, in part because the old prosthesis has to be removed before a new one can be inserted. "But the surgeon and I decided to try again," Wawrzonek said. "That one lasted two weeks." When it failed, they decided to try a third time. "That was the disaster." The moment the third operation ended, he said, he knew that his right ear was dead. "There wasn't a trace of any sound, even after they'd taken everything out." He told me that he holds the surgeon 50 percent responsible for the loss of his right ear, and himself 50 percent responsible. "Doctors should be conservative, and he wasn't conservative," he said. "And either one of us should have had enough brains not to do it. Just say, Whoops, you've still got some hearing there, get a hearing aid and leave it alone. But we did it a third time, and when the surgery was finished I knew that it was all over."

Wawrzonek has now lost quite a bit of hearing in his left ear, as well—partly because he's getting older, and partly also because stapedotomies can gradually deteriorate, as repeated movements of the wire that loops over the incus cause some of that bone to erode or necrotize. He wears a powerful hearing aid, and for our conversation he positioned himself with his back toward the angle created by a wall of the kitchen and a big window. Although he's very hard of hearing—he sometimes had trouble understanding me if we weren't facing each other—he takes advantage of the hearing he has left. He and his wife bought their house, he said, because of a high-ceilinged room that opens off of the breakfast room. The acoustics in that room are so good that musicians sometimes use it as a recording studio, and when I drove up he himself was playing Gershwin on the piano. And he still enjoys concerts, as long as the venue is suited to his particular disability. Just the week before, he said, he and his wife had attended a performance by the Borromeo String Quartet at the St. Botolph Club in Boston.

"The room they have for music has fairly dead acoustics, and things sound wonderful to me there—absolutely phenomenal," he said. "They did Bach's Goldberg Variations, and I'd never really appreciated it before. I was sitting twenty feet away from the quartet, so I could hear almost everything except the high frequencies, which I can't hear at all. It's the only place that I can go and hear that well. It was stunning. It was the best musical experience I've ever had, despite listening through a hearing aid and with only one ear."

IN 2012, NADINE DEHGAN, a consultant with nonprofits in New York City, noticed that her one-year-old daughter seemed developmentally delayed in comparison with her sister, who is a year and a half older. "She was missing all kinds of milestones, and she wasn't speaking, wasn't responding, wasn't being social," Dehgan told me. "She would just sit by herself in the corner and build towers and smash them down." Her behavior and responses were within the range of the normal, though only just, Dehgan said, and if she and her husband, a mathematician and quantitative analyst, hadn't reviewed videos of their older daughter at the same age they would have been less worried.

Finally, when their younger daughter was eighteen months old, Dehgan shared her concerns with her pediatrician. "I came in saying, 'I think there are issues,' and they said, 'OK, we'll get her tested.'" A New York State agency performed tests for learning disability, autism, cognitive issues, and speech delay. The testers concluded that Dehgan's toddler was learning disabled and that she was far behind other children in her age cohort, with a developmental age somewhere around six months. The Dehgans had her retested by a third party, a private clinic in Tribeca. "They told us that she was autistic and that she needed immediate intervention."

Dehgan adjusted her work schedule so that she could devote herself to accommodating her daughter's needs. "I was sad to hear the diagnosis, but I was glad that we'd caught it early, because early intervention is the key with developmental issues," she told me. "I started researching schools in Boston, because I had heard that that was the best place. And we took her to intense therapy." Because Dehgan was preoccupied with those activities, she was several months late in taking her daughter to the pediatrician for her two-year-old checkup. "When I finally did, they gave her a hearing test—which she failed," Dehgan continued. The cause of her developmental delays wasn't autism; it was hearing loss. "The pediatrician hadn't thought to give her a hearing test when I brought her in initially. And neither the state nor the third party had tested her hearing, either."

All newborns in New York state are supposed to be given a hearing test immediately after birth. Dehgan's daughter had passed hers, and, presumably for that reason, no one, including her parents and the Department of Health, had thought of deafness as a diagnosis. "I consider myself an involved and somewhat educated parent, and so is my husband—yet we both missed it," Dehgan said. "And I even had a younger brother with hearing loss. Which gives me pause." The pediatrician's failure to consider hearing loss as an explanation is harder to understand, because the child had had ear problems almost since birth. "Her ear canals are oddly shaped, it turns out, and she had so many ear infections when she was a baby that her eardrums would rupture, and gunk you don't need to know about was coming out," Dehgan said.

Dehgan's daughter's problem, like Jeannette Barnes's and John Wawrzonek's, was a form of conductive hearing loss. Scar tissue from repeated severe infections had hardened her eardrums and filled her ear canals, so that virtually no vibrations were making it all the way to her (undamaged) cochlea. Luckily, hearing loss of that type is almost always treatable. A surgeon operated on her several times, to remove scar tissue and other blockages, and inserted tubes through her eardrums to drain her middle ears. "Right now, she's doesn't have perfect hearing, but she is managing without hearing aids," Dehgan said. "She's a social, happy kid—and you wouldn't guess that there was a problem. Once sound was allowed into her ears, she was able to make up for lost time, because we really worked with her. And it was caught early, so she was one of the lucky ones." Her hearing is affected by colds and ear infections, but her teachers learned to move her to the front of the classroom and make other accommodations.

Dehgan's brother was less fortunate. "His hearing problem was detected when he was a child, but not a young child," she said. "He had made it through elementary school with limited hearing, and kind of just overcompensated. He stuck to sports and places where he was comfortable, and then, once his hearing loss had been diagnosed, my mom made sure that he would at least leave home with his hearing aids." Dehgan said that there could be a genetic connection between her brother's problem and her daughter's, but that they hadn't been tested for that.

Thomas Edison became increasingly deaf beginning in childhood. He attributed his deficit to an incident that occurred when he was twelve and working on a train. He said that the train's baggage master, to punish him for accidentally starting a fire with some chemistry equipment, "boxed my ears so severely that I got somewhat deaf thereafter." Later in his life, he told a different version of the train story, in which the problem began when a conductor grabbed him by the ears to pull him into a moving boxcar. "I felt something snap inside my head, and my deafness started from that time and has ever since progressed," he wrote. Those incidents made a huge impression on me, more than fifty years ago, when I watched Mickey Rooney reenact both of them in the movie *Young Thomas Edison*. It's not certain that either one really occurred—or, if they did, that they contributed to Edison's difficulties. His deafness most likely had a genetic origin. The near consensus nowadays is that Edison's hearing problems were primarily conductive, and that the likely principal cause was either otosclerosis or mastoiditis, a bacterial infection which is typically caused by severe middle-ear infections that spread beyond the middle ear and, in some cases, harm the auditory system or spread to the brain, causing meningitis. The damage to Edison's hearing may have been aggravated by scarlet fever. "I haven't heard a bird since I was twelve years old," he wrote in 1885.

Edison didn't always view his loss as a loss. "Throughout his life Edison would claim that his poor hearing was an advantage; that it reduced distractions by enabling him to concentrate," Paul Israel, the director of the Thomas A. Edison Papers, at Rutgers, wrote in *Edison: A Life of Invention*, published in 1998. Edison heard reasonably well when he was a young man; later, he said, his deafness allowed him to ignore "the babble of ordinary conversation." When he auditioned musicians for early phonograph recordings, he bit down on a wooden sound board attached to the phonograph. Some of the vibrations passed through the wood, through his jaw, through his temporal bone, and into his cochlea, bypassing his eardrum and middle ear.

This form of sound transmission is known as bone conduction. People with normal hearing rely on it, too; it's part of the way we hear our own voices. (Bone conduction makes us think our voices are lower in pitch and more resonant than they sound to others, and that's the main reason we cringe the first time we hear ourselves speak on a recording.) Bone-conduction devices work only when the inner ear has at least some function. There are bone-conduction hearing aids, intended mainly for people whose principal hearing defects are in their middle ears or external ear canals; the most effective ones are physically anchored to the skull. There are also bone-conduction headphones for people who have normal hearing but want be able to (for example) listen to music or communicate with other people without interfering with their ability to hear ambient sounds. The Navy SEALs who killed Osama bin Laden, in 2011, spoke to each other during the raid with "bone phones," which left their ears uncovered. There are also versions intended for runners, who need to be able to hear cars approaching from behind, and scuba divers.

Edison sometimes argued that his form of deafness had made his inner ears more sensitive than other people's, by protecting them from "the millions of noises that dim the hearing of ears that hear everything." Still, he spent several months trying to invent a hearing aid, believing that the market would be huge. Israel quotes a longtime private secretary's recollection that Edison "felt the loss of his hearing very much when he had visitors, and if they told funny stories among themselves and laughed hilariously, a wistful look came over his face, for he was very fond of humorous stories."

NADINE DEHGAN, because of her experience with her daughter, became involved in raising money for hearing research, and in 2016 she became the chief executive officer of the Hearing Health Foundation, which until 2011 had been known as the Deafness Research Foundation. (She left that job in 2018, but she's remained involved in hearing issues. "I'm now that crazy person on the train," she said. "When I hear kids listening to music too loud, I'm like, 'You're damaging your ears!'") The foundation was created in 1958 by Collette Nicks Ramsey Baker, who had suffered a severe conductive hearing loss in adolescence. Baker later underwent operations on both ears, and said that if the surgery was successful she would make a significant donation in support of hearing research. The foundation was the result.

I found an obituary for Baker online. It said that she had been born in Waverly, Tennessee, in 1918, and had died, at the age of ninety-one, in 2010. It also said that she'd had two daughters, and that one of them, Collette Wynn, lived in the same small town in northwestern Connecticut where my wife and I live—an amazing coincidence. I found Wynn's phone number in our town's (nine-page) directory and called her. She invited me over for tea. We'd never run into each other before, somehow, but we turned out to have friends in common, since we both play bridge.

"Mother was profoundly deaf," she told me. "I was her ears. I would always answer the phone for her, and take notes on conversations, et cetera, et cetera." Her mother, she said, began to lose her hearing when she was thirteen. She was swimming with friends in a pond in Florida, where she lived, and dove into shallow water. Her head struck the bottom, and the impact damaged or dislocated her auditory ossicles, giving her the traumatic equivalent of otosclerosis. She dropped out of school. Her mother had died the year before, and she didn't like her father's second wife, and when she was still a teenager she went to live with an aunt in New York City. "She entered a beauty contest and won it, and that attracted the attention of Howard Chandler Christy, a very famous painter," Wynn said. He is best known for his creation of the Christy Girl, an idealized beauty who appeared, in numerous versions, in magazine and book illustrations in the late 1800s and early 1900s, and, most memorably, in a sexy World War I recruiting poster whose caption is "Gee!! I wish I were a man. I'd join the Navy." Baker became one of Christy's favorite models; she appears prominently, as a naked nymph with long blond hair, in a mural that he painted for Café des Artistes, in New York. "Mother never wanted me or my younger sister to know that she had posed in the nude," Wynn said. "She did tell my middle daughter, with whom she was very close, but she said, 'Don't tell your mother.'"

In 1936, when Baker was eighteen, she met Hobart Cole Ramsey, a wealthy New Jersey industrialist, who was twenty-seven years older than she was. At the time he met Baker, he was engaged to the film star Norma Shearer, but he broke off that relationship and married her instead. "Mother was brilliant, absolutely brilliant, really the smartest woman I've ever met, even though she was embarrassed that she hadn't gone to college," Wynn said. "She was also a raving beauty."

She couldn't hear much, but had become adept at lip-reading. Ramsey asked her to abandon a possible career as an actress, to stop flying airplanes (she'd just earned a pilot's license), and to learn bridge and golf, so that she could play with him. She became highly successful at both games, and, when she was in her mid-twenties, she won the women's club championship at Baltusrol Golf Club, of which Ramsey was the president, four years in a row. "Other women golfers thought she was a snob, because they'd say hello and she wouldn't answer, but she just couldn't hear them," Wynn said.

Baker had the first of two ear operations in 1952, when she was thirty-five. The procedure was fenestration, the immediate predecessor of the stapedectomy, and the surgeon was Julius Lempert, who had refined it. He removed part of the malleus and created a new membrane-covered opening in the wall of one of the semicircular canals, as a sort of alternative oval window. (You can watch Lempert performing a fenestration, in an old film on YouTube. Search for "Fenestration Surgery for Otosclerosis—Historical.") "Afterward, she told me, 'I can hear the grass growing,'" Wynn said. Lempert operated on her other ear two years later, also successfully.

Baker created the foundation shortly after the second operation, with financial backing from her husband. It was an early supporter of the research that led to the development of cochlear implants, and it currently funds three main research programs: the Hearing Restoration Project, which was established in 2011 and is an international consortium of scientists working collaboratively on, among other things, the regeneration of damaged hair cells, with the goal of curing both sensorineural hearing loss and tinnitus; Emerging Research Grants, which awards seed money to young scientists doing early-stage research in hearing and balance disorders; and Ménière's Disease Grants, which the foundation began awarding in 2017. (Among the 2018 recipients of Emerging Research Grants is David Jung, the surgeon I spoke to, who is working on a novel method of delivering potentially therapeutic substances to the inner ear.) Baker also helped to establish the predecessor of the National Temporal Bone, Hearing and Balance Pathology Resource Registry, which provides cadaver specimens to researchers all over the world. The registry is now part of the National Institute on Deafness and Other Communication Disorders and is housed at Mass. Eye and Ear. If you don't mind the thought of pathologists and medical students using a band saw to remove parts of your skull after you're dead, you should consider donating yours.

THE OFFICES OF THE HEARING HEALTH FOUNDATION are on the tenth floor of a nondescript office building near Madison Square Garden, in New York City. The building's location serves as a useful reminder of the principal cause of many of the problems that the foundation has spent six decades addressing: traffic noise from Seventh Avenue is virtually constant and is frequently punctuated by conversation-annihilating sirens and vehicle horns. When I visited, in 2017, Nadine Dehgan said they'd been tormented recently by someone playing a trumpet on a street corner directly below. "People were opening windows and screaming at him to stop—and there's even a police station right around the corner—but he kept at it for hours," she said. Manhattanites tend to become inured to the cacophony of city life, but some annoyances are beyond bearing. In the late 1970s and early 1980s, my wife and I lived in a fourteenth-floor apartment in Manhattan, and I became adept at throwing eggs, underhand, from our living room window at a noisy bar on the far side of Second Avenue. I also found the home phone number of the owner and, until he changed it, woke him up every time his bar woke me up.

At the time I visited, Dehgan and her staff were getting ready for a resource fair that the New York Police Department was about to conduct for the deaf and hard of hearing. The foundation had printed information cards intended to help people who have hearing problems communicate with law-enforcement officers. The HHF's cards were based on more comprehensive ones created by Minnesota's Department of Human Services. The idea is that a deaf person who's been stopped by the police can prevent misunderstandings by presenting one of the cards, which say, among other things: "A hearing aid or cochlear implant does not allow me to understand everything," and "Please do not shine a flashlight in my face; it will make me unable to read your lips and understand you." This is not merely a matter of convenience. People who can't hear what's being said to them sometimes behave in ways that are interpreted as intransigence, or worse, and if the people doing the interpreting are armed, the consequences can be dire. During the early months of the First World War, the British deployed sentries in parts of the country that were thought to be vulnerable to infiltration by spies, and until their training caught up with them they were a menace to people with hearing problems, who didn't always stop when they were told to stop and were therefore sometimes shot.

An ongoing focus of the HHF is a policy change that New York made in 2009, when its Office of School Health discontinued all hearing screening of elementary students. The city, in announcing the change, gave two justifications: "There are no high-quality research trials which demonstrate that hearing screening in this age group leads to better functional or educational outcomes," and "The vast majority of children who fail a hearing screen have hearing loss due to fluid in the middle ear or wax in the external ear canal. These are temporary conditions."

But all this is wrong, Dehgan said. The city's new policy contradicts a statewide screening requirement and also the longtime recommendations of the Centers for Disease Control and Prevention, the American Academy of Audiology, and the American Academy of Pediatrics, all three of which advocate regular, repeated testing. The AAP recommends testing at school entry; at least once at ages six, eight, and ten; at least once during middle school; at least once during high school; and upon the matriculation of any new student who arrives without evidence of a previous screening. It also recommends frequent testing for "students with other known health or learning needs; speech, language, or developmental delays; or a family history of early hearing loss"—all of which can be either causes or symptoms of hearing problems. Studies by the CDC have shown that between birth and first grade the percentage of children with hearing problems increases by two orders of magnitude—from roughly 0.17 per cent to almost 15 per cent. (Estimates vary, in part because measurement methods and definitions of hearing loss vary.) Jeannette Barnes and Gerald Shea both would have been helped by better, more frequent school screening, and if Dehgan's daughter's deafness hadn't been caught when it was, she might have continued to be treated for autism. And even "temporary conditions" can have long-term educational impacts. A young family friend of the Dehgans also had a conductive hearing problem, caused in part by an accumulation of fluid in his middle ear. His condition was also misdiagnosed, initially, as a learning disability. If that mistake hadn't been caught, he likely would have been diverted onto an entirely different educational path—an outcome that is especially common with children who lack ready access to health care or whose families aren't fluent in English.

Laurie Hanin is the executive director of the Center for Hearing and Communication, a nonprofit organization based in New York. She told me, "People don't realize what's not happening. New York City's is the only school system in the state not to have a mandated hearing-screening program." The city's school system is so large, she said, that the state allows it to set its own guidelines, and its decision to end screening closely followed an audit by the comptroller's office that found that existing efforts were not being handled well. "In a way, that's understandable," she continued. "It's a huge school system, with a million kids. But instead of canceling the program, the city should have fixed it. There is now no safety net to catch these kids."

Not all hearing screenings are equally useful. (When my own daughter entered kindergarten, in the small Connecticut town where we live, the test consisted of the school nurse asking her, "Can you see good? Can you hear good?") An economical way to screen even a million children would be to use the same kind of test that's used for newborns. That test measures what are called otoacoustic emissions, which are given off by the inner ear; it involves inserting a tiny microphone into the ear canal. It's noninvasive, it can be administered by people who are not audiologists, and it doesn't require more of its subjects than briefly sitting relatively still. It's not a perfect test, but it quickly identifies kids who are at risk of hearing loss—even those who are too young to answer questions. It's also cost-effective, because hearing-impaired children who are identified early are more likely to learn to read, write, and do arithmetic when they're supposed to, and less likely to fall so far behind their classmates that they require expensive support services throughout their school careers.

"Private schools insist on having children's hearing and vision tested, and for good reasons," Dehgan said. "Not giving all children the intervention they need limits their emotional, social, and academic growth. It's just maddening that there might be kindergartners who aren't being identified. They're labeled as slow or as having developmental problems, when the truth is that they can't hear what's going on. And even kindergarten is too late, because children need those stimuli for language and reading."

But effective hearing exams are easy to perform, and the problems that they are able to identify, if left unnoticed and untreated, are not only disastrous for their sufferers but also, in the long run, ruinously expensive for taxpayers. It makes no sense not to test children as frequently as the AAP recommends.

Schools need to be examined, too. In March 2019, I visited Arline Bronzaft, a retired professor of environmental psychology, in her apartment, in New York City. In 1975, she and a coauthor published an influential research paper that arose essentially by accident. "One of my students, at Lehman College, told me that her child attended an elementary school next to an elevated train line and that the classroom was so loud that the students were unable to learn," she said. The school was PS 98, in Inwood, up near the northern tip of Manhattan, and the track was 220 feet from the building. Bronzaft's student said she and other parents were planning to sue, but Bronzaft, whose husband was a lawyer, told her that if they were going to do that successfully they would need to prove that their children had been harmed. Bronzaft discovered that, in classrooms on the side of the building facing the tracks, passing trains raised decibel readings to front-row rock-concert levels for roughly thirty seconds every four and a half minutes, and that, during those periods, teachers had to either stop teaching or shout, and then had to work to regain their students' attention. She obtained three years' worth of reading-test scores from the principal—"I must say, he was an activist principal"—and was able to demonstrate to the city that, by sixth grade, students on the track side of the building had fallen behind students on the quieter side by eleven months.

"The study got a lot of attention," she told me. It also prompted her to become involved. She helped persuade the MTA to install rubber pads between the rails and the ties on tracks near the school (and, eventually, on tracks throughout the system), and she helped persuade the city to cover classroom ceilings with sound-deadening acoustic tiles. In a follow-up study, published in 1981, she was able to show that those measures had been effective and that the gap in students' test scores between the exposed and less exposed parts of the building had disappeared.

Seven

HEARING AIDS

Thirty-six hundred years ago, Egyptian physicians treated people who had hearing problems by filling their ear canals with mixtures of olive oil, red lead, ant eggs, bat wings, and goat urine. Supposed cures in later centuries included mercury pills; oil made from earthworms or walnut bark; cauterization of the Eustachian tubes; applications of eardrops made from peach kernels fried in hog lard; drilling holes in the skull, to provide alternative "pathways" for sound; fracturing of the mastoid bone with a hammer, to jolt ears into hearing; irritating skin behind the ear to raise pus-filled blisters, in the hope of drawing the deafness away; administering electric shocks; syringing ear canals with boiled urine or with water heated almost to boiling; massaging eardrums with pulses of compressed air; "bleeding from the jugular veins" to reduce "congestion in the finer vessels"; having the patient jump from great heights or gargle while lying down; painting the tonsils with silver nitrate or with a poisonous compound extracted from hellebore roots; praying; exorcism; and opium. Some treatments—placing hot coals in the mouth—arose from a failure to grasp that the silence of the deaf was a result of their inability to hear, not their refusal to speak.

Beethoven's hearing problems began in the 1790s, when he was in his twenties. Doctors at various times treated him by filling his ear canals with cotton soaked in olive oil, strapping rash-causing toxic plants to his forearms, bleeding him with leeches, and sending him to live in the relative silence of the countryside. Something that actually did help was a device that he himself probably created: a rod that he attached to his piano and clenched in his teeth. The rod enabled him to hear his own music, by bone conduction, though only faintly.

Deafness has always been irresistible to scammers. In the late nineteenth and early twentieth centuries, many Americans with hearing problems bought artificial eardrums, thimble-like rubber inserts that were claimed to be "to the ear what glasses are to the eye." One advertisement said they could protect users from being run down by recent innovations in public transportation, since "Rapid Transit requires of the public the keenest hearing of the gong." (This was a genuine safety issue. Electric trolleys moved so much faster than horses that pedestrians who weren't accustomed to jumping out of the way were sometimes run over—a hazard that was the source of the original name of Brooklyn's baseball team, later relocated to Los Angeles: the Trolley Dodgers.) Charles Lindbergh was one of a number of early aviators who took people with hearing problems on "deaf flights," in which tailspins, nose-dives, barrel rolls, and other stunts were supposed to force defective auditory systems into recovery. Not all the crazy-sounding treatments were entirely crazy; a nineteenth-century Englishman restored at least some of his lost hearing by covering his ruptured eardrum with a small piece of moistened paper, creating a crudely functional temporary patch. And quack treatments are less common today than they used to be, although I periodically receive email from someone who claims to have cured his wife's deafness with an herbal concoction whose recipe was revealed to his mother by a "Navajo medicine man." If you bite, for thirty-seven bucks, you get the story, not the herbs, because, he explains, if he actually revealed his secret the global hearing-aid cartel would destroy him.

The first hearing aids were artificial versions of cupped hands, and there's evidence that ancient Egyptians and Greeks used hollowed-out animal horns in that way. In the late 1600s, an English baronet invented a number of "speaking trumpets," which were used to amplify the voices of church choirmasters and others. Ear trumpets were fairly common a century later. A prominent user was the painter Joshua Reynolds, to whom James Boswell dedicated his *Life of Samuel Johnson*. Reynolds's hearing problems began during a trip to Italy and were attributed by him to the lingering effects of a bad cold. In 1819, Frederic C. Rein, an English manufacturer of hearing devices, created an "acoustic throne" for the king of Portugal, who had been hard of hearing since early childhood. Suppliants knelt before the king and spoke into the open mouths of lion heads carved into ends of the throne's arms; the lion heads were connected to various conduits and, eventually, to a tube that the king held to his ear. Rein's company, in London, manufactured hearing devices until 1963.

The first electric hearing aids were introduced in the early years of the twentieth century. They were expensive, cumbersome, and enormous, in addition to not working very well. Many were "table-top" devices. In 1921, Western Electric began manufacturing the Vactuphone, which weighted thirty-five pounds and came with a single Bakelite earphone nearly as large as a cupcake. For decades, hearing-aid components had to be strapped to the body or carried in sturdy pockets. In the 1940s, Zenith suggested that women users of its Sonotone hearing aid wear the transmitter (roughly the size of a pack of cigarettes) "in the hollow of the bust," and strap the battery either under an armpit or between the top of a stocking and the bottom of the girdle, with wires concealed under other garments. The company also suggested that school-age children wear the components in a "special vest," patterns for which it would provide at no charge, or to attach them to a Sam Browne belt—a wide belt with a skinny, diagonal shoulder strap, invented by a British soldier who had lost an arm.

The invention of the transistor, at Bell Labs in the late 1940s, and consequent reductions in the power requirements of many types of electronic devices, eventually made it possible to manufacture "one-piece" hearing aids, which were small enough to be worn behind the ear or concealed in the stems of eyeglasses. My earliest memory of my grandmother's hearing aid is probably from the 1960s. The device had a little dial for controlling the volume, and it was so bulky and angular that she had a hard time positioning it. The batteries were small by comparison with flashlight batteries, though far larger than hearing-aid batteries are today. Occasionally, in the middle of a conversation, she would realize that she couldn't hear what people were saying and guess that her battery must be dead; she would then rummage through her purse.

TODAY, MOST OF THE WORLD'S HEARING AIDS are made by six manufacturers, just one of which is based in the United States: Starkey
Hearing Technologies, a privately owned company, whose headquarters are in Eden Prairie, Minnesota. (The other five are Phonak,
in Switzerland; Signia, in Singapore; and Oticon, ReSound, and Widex,
all in Denmark.) I visited Starkey in March 2017. When I arrived at
the testing department, the receptionist greeted me in a voice she
seemed to have turned up—an occupational necessity, I assumed. I
sat in a waiting area with other people who had come to have their
ears tested. Most of them were older, but one was quite a bit younger.
An employee told me, "Where you're sitting have been astronauts,
royalty, and presidents"—and photographs of many celebrated patients were hanging on the walls. She said that I had been preceded
recently by two members of a well-known rock band, which has been
around since the early 1970s. A big problem for rock musicians is that,
in order to play, they have to be able hear their own instruments over
everyone else's. In the old days, they managed that by placing wedge-
shaped speakers at their feet, aimed at themselves, and turning up the
volume when they couldn't hear them anymore: an arms race. Michael
Santucci—whose company, Sensaphonics, makes an extensive line of
in-ear monitors and protective devices for musicians, and who is known
as "the audiologist to the rock stars"—told me that the members of
one world-famous band used so many wedges onstage during one of
their tours that one of them suffered vestibular symptoms if he stood in
certain places. The rockers who preceded me at Starkey had ignored
their hearing problems for decades. "They looked very old and very

weathered," the Starkey employee said. "Oh, my gosh, they've lived hard. But they're still playing." And they have hearing aids at last.

When it was my turn, a technician used a digital otoscope and a curette—a long, wirelike tool with a scoop at the end—to remove wax from my auditory canals. I'd felt slightly nervous, the night before, about showing up with embarrassingly waxy ears but now realized that if anybody was used to seeing earwax, it was this guy. I watched what he was doing on a wall-mounted video monitor, which magnified everything forty times and made the curette look like something you might see on a construction site. A typical ear canal is sort of S-shaped, with a little twist at the end. Each one is as individual as a fingerprint, he said. I saw many pale hairs (not hair cells), and my eardrum, and pieces of wax that, under magnification, seemed almost minivan-size. He wiped the bits of wax on a napkin he'd placed on my shoulder. He pointed out several bumps on the walls of my ear canals and said they were benign bones growths, called exostoses. He explained that they usually form in response to repeated exposure to cold water and are the ear's attempt to insulate its vulnerable interior parts. Exostoses are common among cold-water surfers (who refer to the condition as "surfer's ear") and among swimmers. Large exostoses can cause conductive hearing loss, by blocking or partially blocking the auditory canal. My *New Yorker* colleague William Finnegan—whose surfing memoir *Barbarian Days* won the Pulitzer Prize for biography in 2016—has lost some hearing to exostoses and has been operated on for them. I've never surfed, although during the summer I do swim in an unheated pool and during the winter I sometimes play golf in the snow. "But yours are too small to worry about," the technician said.

For my ear test, I sat on a chair in a small sound-isolation booth. An audiologist inserted receivers into my ear canals, then went outside, closed the door, and played tones that I could hear through my headset. She had told me to press a button every time I heard a tone, no matter how faint. As soon as I couldn't hear anything anymore, she spoke words into my headset and asked me to repeat them. Then she spoke to me at different volumes and asked me to choose the level that felt the most comfortable. Then she played pairs of piercing, high-frequency tones, and, in each case, asked me to indicate which tone sounded more like my tinnitus—and in this way she was able to pinpoint my tinnitus at around 6,000 hertz.

When we'd finished, she showed me the results on a chart: my audiogram. It depicted my hearing "thresholds," the quietest sound levels I could discern, measured in decibels, at regular intervals across a range of frequencies. The faintest sound that a human with perfectly functioning ears can hear half the time is designated 0 dB; as our ability to hear declines, more decibels are needed to penetrate our awareness at each frequency, and when that happens our thresholds are said to become "elevated"—a slightly confusing term because on the audiogram of someone who has lost hearing, the line on the graph bends down, not up. As a practical matter, hardly anyone older than grade-school age can hear sounds close to the bottom of the scale (although Mark, the hyperacusis sufferer, can now hear sounds so faint that many newborns wouldn't notice them). Adults whose thresholds are elevated 20 dB are still usually classified as having normal hearing. A mild loss is usually described as an elevation between 20 dB and 40 dB; moderate, between 40 dB and 70 dB; severe, between 70 dB and 90 dB; profound, 90 dB and beyond. Someone with moderate hearing loss can have trouble with normal conversations even in a quiet room; someone with profound hearing loss can't hear much at all.

My hearing test showed that in the lower frequencies my thresholds in both ears are fairly close to the middle of the normal range, with elevations of around 10 dB, but that my hearing begins to decline at around 2,000 hertz, which is roughly the frequency of the highest note you can play on a flute. It doesn't move out of the normal range until about 4,000 hertz, which is somewhat lower than the highest note you can play on a piano. It remains in the mild-loss band until almost all the way to the right-hand side of the chart, at 8,000 hertz, where my left ear creeps into the moderate zone. According to the audiogram, my left ear works slightly better than my right at lower frequencies but slightly less well above 2,000 hertz.

My results were not atypical for a non-rocker in his early sixties; in most people with noise-induced hearing loss, it's the higher frequencies that go first. In many cases, thresholds are sharply elevated at around 4,000 hertz, and then improve slightly at 6,000 or 8,000 hertz—a pattern known as a "notch" because the line on their audiogram dips down in that region and then moves back up. (Such a dip has sometimes been called "boilermaker's notch," after its earliest identified sufferers.) Deficits in the middle and high frequencies are among the reasons that people in middle age dislike going to noisy bars, crowded dance clubs, and big parties filled with boisterous young people: as the ability to hear higher frequencies fades, speech becomes harder to understand. This is largely because consonants are pitched higher than vowels, and when consonants disappear sentences turn to mush—making "face the facts" difficult to distinguish from "fake the stats." Speaking louder to someone who's having that kind of hearing trouble seldom helps much, because merely increasing the volume doesn't cause the ear to pick up sounds it can't hear at all. (Hawaiian might be a useful language for people with mild hearing problems: all those vowels.) Some people's voices are easier to hear than others'. The voice of my old golf buddy Harry was very low, and, because low-frequency sounds travel farther than high-frequency sounds, it carried for tremendous distances. I once overheard him asking one of his playing partners a question to which I knew the answer, so I answered it myself, by shouting from two fairways away. "How did he know what I was saying?" Harry asked his playing partner, whose voice I couldn't hear at all.

Most women's voices are pitched much, much higher than Harry's—and that may partly explain a common complaint that oldish wives make about their oldish husbands. I once read about a woman who, when she speaks to her hard-of-hearing grandfather, uses what she calls her "grandpa voice," which she consciously lowers into a range that he can hear. Hearing loss that's concentrated in the lower frequencies is uncommon but does exist. My golf buddy who complained when he thought we hadn't praised his almost-a-hole-in-one has that kind—and that may explain why it took him so long to get hearing aids: he had always been able to hear his wife well enough to get by and, by comparison with that, not hearing his boring male friends didn't seem like a big deal.

Because my hearing loss, though measurable, was too mild to be a genuine nuisance, I'm pretty sure that under ordinary circumstances the audiologist would not have suggested hearing aids. But I was doing research, so I was fitted for a pair, from a Starkey line called Muse. Each unit sits behind an ear, as my grandmother's hearing aid did, but is much smaller. It also comes in more colors. The woman who fitted me had a large tray of possible choices, and selected gray, to match my hair. "If a man is bald, you can sometimes try to match the glasses," she said. A coated wire leads to a receiver— red for right, blue for left. (Another Starkey employee told me, "Even after forty years, I have to remind myself: *right* and *red* start with *r*.") Each receiver is about half an inch long and the diameter of a kitchen match, and it goes right into my ear canal. The receivers she gave me had small tips, which don't fill the ear canal completely. The small tips, she said, would keep me from having a "head in a barrel" sensation, caused partly by the fact that if you completely plug an ear canal, sounds that make their way inside it bounce around, since they have no escape path. The bouncing around is called "the occlusion effect."

Once I had my aids, the audiologist took me to her office and adjusted them, by doing something I couldn't see on a desktop computer. She added two subtle tones meant to mask my tinnitus; I could select one or the other, or neither, by pushing a button on the unit behind my ear. She showed me how to clean my aids, with a little toothbrush, and how to change the batteries, which are a fraction of the size of the ones my grandmother used. The audiologist gave me a case, a storage jar partly filled with desiccant, and a container of what looked like orange toothpicks but are actually applicators for disposable earwax filters. My main reaction when I first put on the hearing aids was mild annoyance at the sound of my voice. I also became more aware of turning pages, creaking doors, and the surprisingly varied noises made by my pants. The audiologist said that getting used to hearing aids can take a month, as the brain adjusts to receiving unfamiliar inputs, and that, for that reason, it's important not to give up.

For people who wait longer than they should have before doing something about their hearing loss, the improvement can be dramatic. A hard-of-hearing friend who waited to be fitted for hearing aids until he was seventy told me in an email: "I went last night with two brothers and a sister-in-law to a very noisy bar. It was my first test of my hearing aids in noisy conditions. I went in with one brother, and as we were waiting for the other brother and his wife I could hear him as well as anyone ever can in that environment, so I had to ask him whether he was using a speaking-to-a-deaf-guy voice or just a speaking-in-a-noisy-bar voice. He assured me it was just the latter. Then when the other brother arrived I was again so thunderstruck by how clearly I could hear that I had to ask him the same question. Same answer!"

In 2018, a reddit user asked hard-of-hearing reddit users what had surprised them the most when they first got hearing aids or cochlear implants. Among the answers: farts; toilet flushes; peeing; refrigerators; the fact that sunlight doesn't make a sound; the fact that falling rain makes a sound but falling snow does not; the annoying loudness of typing and other routine office activities; cloth rubbing against cloth; hair brushing against hearing-aid microphones; cutlery scraping on dinner plates; clocks ticking; the silence of sharks; the relative silence of cabinet hinges; that vocal intonation can be used to distinguish sincerity from sarcasm; that fire doesn't sound like a continuous explosion; that voices don't all sound the same; that songs have intelligible lyrics; that music is more than its bass line; that grocery stores play background music; and "What's weird is, boobs don't make a noise, you really think they would."

On the other hand, people who suddenly gain the ability to hear sometimes miss the tranquility of deafness. A woman who responded to the same reddit question said that her grandmother, after receiving bilateral cochlear implants, "was always really pissed off about the frogs in her pond because they never shut up and she could no longer sit and read by the pond." Another respondent: "My wife refused to get hearing aids for years. We bought a house that backs on a four-lane highway. I tolerate noise well. When she came home with her new hearing aids, she went out to the backyard, came back and asked me 'WHY DID YOU LET US BUY THIS HOUSE?'"

STARKEY'S CHAIRMAN AND PRINCIPAL OWNER, William F. Austin, was born in Nixa, Missouri, in 1942. He attended the University of Minnesota Medical School and, on the side, worked in a small hearing-aid repair shop owned by an uncle. He has said that, while working in the shop, he decided that practicing medicine would be less useful to humanity than helping people hear better, so he dropped out. In 1970, he paid $13,000 for Starkey Labs, a small company that manufactured earmolds, which are plastic hearing-aid components that fit inside the ear canal. A few months later, he began making hearing aids of his own, beginning with a model that he called the Custom Master.

Austin was an intuitive salesman, and in the company's early years he introduced several innovations that were copied later by other manufacturers, among them a ninety-day free-trial period and a one-year full warranty. His greatest marketing triumph occurred in 1983, when President Reagan revealed that he was wearing a Starkey hearing aid. "A White House official said Mr. Reagan had already developed the habit of using and then removing the hearing aid at will, much like a pair of glasses," the *New York Times* reported at the time. "The official said the President had told aides he intended to use it mostly for meetings at the White House." The main source of Reagan's hearing problem was apparently a .38-caliber pistol that someone fired near his right ear while he was making a movie in the 1930s. Reagan's ear doctor, John William House—whose brother, Howard, established the House Clinic, which is still one of the country's most distinguished otologic practices, and the House Ear Institute, a major research organization—said, "Maybe the President's doing this will help others realize that they can ease their problems with a hearing aid." That did happen. Starkey's sales doubled almost immediately, and sales by other manufacturers rose, too. Austin said later, "In no time, we were buried in orders. We had hearing aids stacked to the ceiling." The company rapidly expanded into other countries.

By comparison with modern hearing aids, Ronald Reagan's first one was technologically primitive, as were all hearing aids in that era. Improvements in transistors and batteries had made hearing aids easier to conceal and more comfortable to wear, but the aids were analog devices, like a phonograph, and they didn't do much more than make sounds louder. Advances in semiconductors and digital signal processing created new possibilities, and, beginning in the mid-1990s, as microprocessors rapidly increased in power while shrinking in size, two manufacturers introduced the first digital hearing aids. Among the advantages of those devices was that they could manipulate sound in more complex ways than analog hearing aids could, by making weak signals clearer and by adding precise amounts of gain in the frequency ranges where it was needed.

As this technological revolution was beginning, Dianne Van Tasell, a professor of hearing science at the University of Minnesota, was thinking about making a career change. She was teaching courses in hearing aids and hearing-aid signal processing, and, she told me, "I got tired of people talking about stuff but not doing it." She made an appointment with Starkey's president, Jerry Ruzicka, whom she knew slightly, and told him that the company needed to adopt the new digital technology. "Starkey and other companies were happily making analog devices," she told me. "And they had invested a lot in the tools and equipment to manufacture those devices. Starkey was one of the companies where the feeling was, 'Oh, well, this digital thing is just a passing fancy.' So I told Jerry that Starkey had made the wrong bet, and that they were about to get their butt kicked by everyone else, and that, if they would hire me and allow me to hire the people I wanted to hire, we would develop a digital aid in-house." He agreed. Van Tasell went to work at Starkey full-time.

"Of course, I had no idea how hard that was going to be," she continued. "Hearing aids, technologically, are a marvel. When I first started at Starkey, we went to Texas Instruments and told them that we were looking for a digital hearing-aid solution, and that we knew they made chips by the zillions, and that we'd like them to customize an off-the-shelf chip for us. They were happy to do that— but then we told them what our size and power requirements were, and they said, 'What? You've got to be kidding!'" The challenge with hearing aids is that every component has to be tiny, yet has to function ten hours a day on current drawn from a miniature battery that users expect to last at least a week. "They said that there was no chip that could do that," Van Tasell continued. "And that's why the hearing-aid companies have all invested millions and millions and millions of dollars in the development of their own low-power chips. And it's why they know how to do low-power stuff better than anybody." Developing Starkey's first digital hearing aid ended up taking five years, and when that project was complete, Van Tasell left Starkey—partly because she had accomplished what she'd set out to accomplish and partly because she'd become disenchanted with the company.

AS SOON AS MY HEARING AIDS HAD BEEN ADJUSTED, I was given a tour of Starkey's production area, which looked less like a factory than an ordinary office, with lots of people sitting in front of computer monitors on desks in waist-high cubicles. My guide was Bruce Swenson, who had worked at Starkey for forty years. "Production actually begins down in the far corner of the room," Swenson said. So that's where we started.

My hearing aids are "off-the-shelf"; the person who assembled them for me selected a tip from a tray that contained lots of ready-made ones, in a range of sizes. By contrast, hearing devices that entirely fill the ear canal, as many do, especially ones for people with more severe hearing losses, are custom-made from silicone impressions that are created by injecting goop into patients' ears. The impressions look like pocket-size Henry Moore sculptures or like compact piles of squeezed-out toothpaste: no two ear canals are exactly alike, and none of them are remotely cylindrical. People with unusual ear canals sometimes travel to Minneapolis to have their impressions made right at the factory; FedEx delivers the rest, each morning, from audiologists' offices around the world. The orders are color-coded based on the day they arrive, and Starkey's fabricators work around the clock, in three shifts, with the goal of turning around every order in four days or less.

Hearing-aid makers used to work directly from physical impressions. "We would start cutting, grinding, modifying, detailing, dipping in coating materials," Swenson said. "It was a labor-intensive process, and it was subjective. It never turned out quite the same two times in a row, and if your dog ate your hearing aid you had to send in another impression." Silicone yielded more consistent results than earlier injection materials, which shrank over time, but there was still a lot of variability.

Now all the shaping is done digitally. (There are systems for making the impressions digitally, too, but they're still expensive to use, they aren't perfectly reliable, and they haven't replaced silicone yet.) Each casting is mounted on a pedestal the size of a milk-bottle top and placed inside a three-dimensional laser scanner: a smallish metal box with a computer monitor on top. "It spins, drops, spins, drops," Swenson said. "The device scans both ear impressions at the same time, and in about two minutes we have digital copies." Starkey began exploring a switch to digital modeling in 1991; implementing a working system took another dozen years. "Nowadays, everyone has heard of 3-D printing, but this was like *Star Wars* at the time," Swenson said. "When we made the transition, we retrained our hand-pour technicians to do exactly the same job, using exactly the same skill set, but in a software program. So now they go home without dust on their clothes." Starkey stores the silicone impressions, as emergency references, but all the work is done on the digital copies. The images are trimmed, shaped, and manipulated on a computer program that's essentially Photoshop for ear canals. The goal is to create a hearing-aid shell that will not only fit comfortably but also accommodate all the electronic components that have to be packed into it.

I stood behind another technician as he rotated a scanned image of someone's ear canal on his computer screen. "We have different sculpt tools, and we can fix or repair imperfections," he said. Air bubbles have to be filled in, as do areas where the silicone didn't reach all the way to the edges of the canal. "Most flaws are easy to spot. Where silicone has touched the ear surface, we can see the skin structure—the pores, like here—but where it has touched only air we see a shiny, glassy finish. Also, ears rarely ever have a straight line in them—like, never. So if we see a straight line and a quick drop, we know the impression must have missed something." The technician used a virtual tool, which he'd selected from a ribbon at the edge of his screen, to fill a depression left by an air bubble, and then he used another tool to create a path for a duct that would act as a vent. With his mouse, he positioned a box-shaped template that represented the required electronics. He found room for it, but only just, by tilting it slightly. He clicked on an icon that brought up an image of an orientation axis, like a three-dimensional compass, which allowed him to make sure everything was pointing in the right direction. "Now I'll detail the canal tip," he said. He used a carving tool to shave some material off the end, to keep it from banging into the eardrum. "We basically build the hearing aid in the virtual world before we go out and build it in the real world," he said.

When the technician had finished, he saved the file he'd been working on and forwarded it to production, so that the digital shell could be turned into a physical one. Swenson showed me the room where that happens. It looked like an office copy center: fluorescent lights, beige walls, beige floor, and beige machines the size of refrigerators. The machines were 3-D printers manufactured by a company in South Carolina. A yellow-and-black warning tape was affixed to the floor about a foot in front of them, telling visitors to stand back: if you bump into one of the machines while it's operating, you risk ruining an entire batch of hearing-aid shells.

In each machine, a squiggle of blue light was dancing on the surface of a vat of clear liquid. The light was an ultraviolet laser beam, Swenson explained, and the liquid was a resin that hardened when exposed to it. (I have a handyman version of the same kind of resin at home, in a pen-like dispenser that has a ultraviolet bulb at one end. I recently used the resin to seal some hairline cracks in the bowl of my wife's food processor.) The laser was creating the hearing-aid shells from the bottom up, cross section by cross section, one ultra-thin layer at a time. There were fifteen shells in the machine we were watching, and Swenson said they were about 60 percent finished. After the laser had made one complete pass across the vat, the platform descended slightly. Then a wiper crossed the surface of the resin to remove air bubbles and rewet the uppermost surface of the forming shells, and the laser danced across again. "The laser makes between four hundred and four hundred and fifty passes," Swenson said. "When the shells are finished, the platform will rise back up out of the resin. Each batch takes about an hour and a half." (When I visited, Starkey was planning to upgrade to machines that used LEDs rather than lasers, a change that would bring down costs and make it possible to create shells in colors other than clear.)

The first 3-D printer I ever saw was in the headquarters of a British engineering firm in London in 2006. The engineers used it to make physical models of structures they were working on, and they liked playing with it so much that the firm had had to adopt a lot of rules. The main industrial use for 3-D printing is so-called rapid prototyping: you're trying to create a new product, and you need a part that you can't just find in a drawer or order from Amazon, so you draw it in a computer-aided-design program and bang it out without leaving your desk. The prices have come down so far that companies with 3-D printers don't need as many rules as they used to—and you can even buy home models for less than the cost of a lousy TV—but 3-D printing is still an expensive way to manufacture anything large or complicated, or anything you need more than a few of. For making bespoke hearing aids, though, it's ideal.

"The printing technology we use was used first in dentistry, to make dentures and crowns," Swenson said. "Historically, the hearing industry has let the dental industry go first. They take care of all the biocompatible testing, and then we borrow the application." Dental appliances and hearing aids have similar requirements. They both have to be small, and they have to be shaped precisely and be biologically neutral, and they have to survive inside human orifices.

Once the shells are finished, some of the company's most

experienced workers—many of them with more than twenty years on the job—do the final shaping and install the electronics. Their supervisor told me, "They take it to a surfacing wheel and make it as shallow as possible, because nobody wants to see their hearing aid, right?" The most finicky job is that of the "casers," who install the internal components and connect all the wiring. "They actually plug a listening piece into the receiver," he said. "Then they move the receiver back and forth, ever so slightly, looking for exactly the right locations, so that we don't have any mechanical feedback or mechanical whistling."

For many people with hearing loss, a useful hearing-aid component, available on almost all but very small models, is a telecoil, or T-coil—a small copper antenna that acts as an independent sound source in certain situations. T-coils were first added to hearing aids to make it easier for wearers to use compatible telephones: when the T-coil was switched on, the hearing aid's microphone was bypassed in favor of an electromagnetic signal received directly from the phone. Today, T-coils work with additional devices, including many television sets, and also in public spaces that have been equipped with audio induction loops, which are antennas that transmit directly to T-coil-equipped devices. If you're watching a play in a looped theater, you can turn on the T-coil in your hearing aid (unless it's set up to switch automatically) and clearly hear only what's happening on-stage, not the person coughing or whispering two rows ahead. And if you don't have hearing aids (or if you have them but left them at home), you can usually rent headphones that connect to the same system. I have a hard-of-hearing friend who loves the fidelity of the headphones, and uses them in preference to his T-coil-equipped hearing aids when he goes to the theater.

LATER THAT DAY, I was given a tour of Starkey's main research building by Jason and Elizabeth Galster—one of a number of married couples working at the company. Jason was the company's senior manager of audiology research, and Elizabeth was a research audiologist. They met not at Starkey but at Vanderbilt, where both earned graduate degrees. Among their common interests is tinnitus: Jason has it, and Elizabeth helped to develop Starkey products that mask it. One of those products is a smartphone app called Starkey Relax, which I sometimes use when I'm working. It features a library of a dozen masking sounds: Chimes, Rainforest, Ocean Waves, Rainfall, Marimba, Acoustic Guitar, Babbling Brook, Thunderstorm, Nature, Oscillating Fan, Crackling Fire, and Starkey Relief Sound (a form of white noise, similar to a tinnitus-relief option available on Starkey hearing aids). All of them are adjustable in volume, pitch, and rate of fluctuation, and if you get tired of one, you can scroll to another. You can also create your own sound files. It's pretty slick.

"Part of what brought me to audiology and the work we do is that I've always had this fascination with sound," Jason said. "When I was an irresponsible sixteen-year-old, I replaced my car stereo with a bigger one, and then with a bigger one, and then with one that was bigger still. I'm sure you've had the experience of hearing a car driving down the street from half a mile away. Well, for some reason, when I was sixteen I thought that was really cool, and I battered my ears." That was roughly twenty-five years ago. "Today, my hearing thresholds are pretty normal, but I have a constant ringing in my ears." He sleeps with a fan.

The Galsters led me past a long wall covered with copies of patents awarded to Starkey employees, themselves included, and we looked into a huge, high-ceilinged room filled with busy people. "This is a modern version of a university research lab," Jason said. "When I was in college, we had a setup just like this, except that we had stacks and stacks of technology that lined the walls, and we've advanced to a state now where we've packaged all of that technology into the desktop computers."

We stopped at a room that contained an office chair surrounded by a dozen loudspeakers on pedestals, which were arranged in a circle maybe twelve feet in diameter. The walls were covered with sound-deadening panels. "This is a setup that we would bring one of our patients to," he said. "We can use these speakers to simulate very complex noisy environments, and we'll ask people to understand speech or complete different tasks in dynamic listening situations." Mounted on the far wall was a video screen, which is ordinarily used to give instructions to test subjects. At the moment, it was covered with a sheet of sound-deadening foam. Jason said, "We have extremely fine acoustic control over everything we do, and, for some testing, the acoustic reflections that bounce off the television screen are too strong, so we have to pad it."

In another room, hearing aids were undergoing "accelerated aging," by being subjected to stresses that were meant to replicate a variety of hostile environments, both inside and outside the ear canal: lengthy exposure to blowing clouds of dustlike talc; baking in an oven suffused with "salt fog"; submersion for days at the bottom of a meter-tall column of water. In the back of the room was a scanning electron microscope, which is used to look for damage caused by the tests. "We can age a hearing aid five years in twenty-eight days," Elizabeth said. Another Starkey employee told me, later, that her father had gotten Muse hearing aids, like mine, and that one day when he was still getting used to them he accidentally wore them into the shower. When he realized what he'd done, he panicked and threw them out of the shower—but into the toilet. He fished them out, and they were fine.

A growing number of modern hearing aids contain Bluetooth wireless technology, which makes it possible to use them instead of headphones for doing things like listening to music, watching TV, and answering the phone. My sister's hearing aids were among the first with that capability. To make Bluetooth work, though, she has to wear a small device around her neck that receives the audio signals from whatever device she wants to listen to and relays them to her hearing aids. The necklace is necessary because the version of Bluetooth used in hearing aids is not very robust. "Hearing aids are orders of magnitude weaker than any other wireless device you'll ever run into," Galster said—a consequence of the fact that the entire device has to run for a week on a 1.45-volt battery the size of an aspirin tablet. And, especially at such low power levels, the head and body act like sponges for the radio waves, blocking communication between (for example) the hearing aid in your left ear and the iPhone in your right pocket.

More recent high-end Bluetooth hearing aids don't need the necklace. Jason showed me the internal antenna from one model. It was made of loops of copper foil and looked a like a metallic butterfly.

"It wraps the interior of the hearing aid, so that you have a right and a left side," he said. "It's a good example of why we have to develop this stuff in-house, because if we didn't it wouldn't exist." In another part of the research building, he showed me a big, boxlike apparatus that had been used in developing the antenna: a radio-frequency anechoic chamber, which, from the outside, looked like a walk-in freezer. Inside, its walls, floor, and ceiling were covered with tightly spaced pyramid-shaped cones made of carbon-based foam, a little like supersized egg cartons. The purpose of the cones was to absorb radio signals emitted by any transmitting device operating within the chamber, to prevent the signals from echoing off the walls, ceiling, and floor. "I call this the Stargate," he said.

I stepped inside. The chamber was eerily hushed, because the foam cones absorb sound waves, too. Near the center, mounted on a human-height stand, was a yellow plastic head, with ears—like a bust on a pedestal. The head's nickname was Homer. "It contains a gel that mimics the frequency characteristics of a human head," Jason said. Homer was surrounded by lasers, which are used to align it, and a ring of sensors. A technician outside the chamber had been bombarding Homer with radio waves, to see which signals were being blocked and which were making it through. Four large monitors outside the chamber displayed colorful blobs of various shapes. Jason said, "If you look here, you see what we're measuring." He pointed to a monitor on which the blob was virtually round, showing that all the signals were reaching their targets. "This, essentially, is what you would get if the hearing aid were floating in space," he said. The other three monitors showed blobs that were deformed in ways that indicated varying degrees of interference from Homer. This is important to measure, he said, because, when you're trying to transmit a Bluetooth signal across the body, "the head becomes a problem."

SOME OF THE RESEARCH DONE AT STARKEY, Jason told me, involves understanding which hearing problems can be addressed by hearing aids and which can't. One area of interest is the extent to which people exploit visual cues in understanding speech, and he said studies have shown that even people with good hearing often rely on unconscious lipreading, which can contribute as much as 20 percent of comprehension. To demonstrate, he covered his mouth with a sheet of paper. "If you can't get those visual clues, listening becomes more challenging and more effortful, even for something like this," he said. "And if you get those visual clues back"—he uncovered his mouth—"you relax."

To pick up clues like these, of course, a listener has to be able to see the speaker's mouth. My friend Patty Marx told me: "Paul and I have a rule in our apartment that you can only talk to someone who is in your same zone. So, for instance, I can't yell at Paul in the kitchen if I'm in the bedroom, because those rooms are in different zones— and if there's a violation you can quietly say, 'You're not in my zone.' The problem is that we disagree about where the borders of the zones are. Also, we always break the rule." My sister had a similar rule when her kids were little: no child sitting in the way-back of her minivan was allowed to speak to her.

The auditory system interacts with the visual system to a surprising degree, and in certain situations the brain allows information from the eyes to override information from the ears. In 1976, Harry

McGurk and John MacDonald, British researchers who were studying how infants learn to speak, accidentally discovered that they
could change their own perception of a spoken sound by overdubbing
video of a speaker making a different sound. If you Google "Try the
McGurk Effect," you can watch a BBC story in which Lawrence
Rosenblum, a professor of psychology at the University of California,
Riverside, says "Bah, bah, bah"—and you hear "Bah, bah, bah." But
then the same audio clip plays over video in which Rosenblum is
saying "Fah, fah, fah," and now what you hear is "Fah, fah, fah," even
though the sound track hasn't changed. Unlike many perceptual illu-
sions, this one doesn't disappear once you know the trick. Rosenblum
says, "I've been studying the McGurk effect for twenty-five years
now, and I've been the face in the stimuli, I've seen stimuli thousands
and thousands of times, but the effect still works on me. I can't help
it." In one part of the BBC story, Rosenblum is shown in split screen,
and whether you hear him saying "Bah" or "Fah" depends on which
half of the screen you happen to be looking at. "The speech brain just
takes in that information and doesn't care what outside knowledge
you bring to bear," he says. The YouTube comments on the BBC story
include one from "Bozeman42," who writes that he now realizes that
he experienced the McGurk effect, repeatedly, while watching and
rewatching the movie *Dr. Strangelove*. President Kennedy was assassi-
nated shortly before the movie was completed, and in post-production
Stanley Kubrick had Slim Pickens dub the word "Vegas" over the word
"Dallas" in one of his lines. "I always heard 'a pretty good weekend in
Degas' and thought he messed up his line," Bozeman42 writes. "My
mind is blown."

Context also plays a role in hearing—a bigger role than most people suspect. Even if a room is fairly quiet and you can see the mouth of the person who's talking to you, your comprehension is probably better if you have at least a rough idea, in advance, of what the person is likely to be saying. Non sequiturs and sudden topic changes make speech harder to understand. An audiologist at Mass. Eye and Ear played me an audio file of what seemed to be pure static. "I can show you scientific research that proves you understood none of that," he said. Then he played an undistorted version of the same file, and I realized that what I'd heard was actually the voice of Homer Simpson—and now, when he played the distorted version again, I could make out everything Homer was saying: "Geez, that dog has more education than I do. He's some kind of superdog!"

"Your brain is your best organ of hearing," the audiologist explained. As hearing ability declines, the brain has to work harder. "One thing we can do with people who've lost some hearing is teach them little tricks," he continued. "So, instead of saying 'What?' you say, 'O.K., great. I'll meet you at two o'clock.' And I say, 'No, I said three o'clock.' So you got the time wrong, but by rephrasing the statement in that way, you were able to fill in just the bit you didn't get— and that seems more natural than saying 'Huh?'"

The audiologist played another recording. This one sounded like distorted speech, rather than just static, but I still understood none of it. He said, "I'm going to give you one word: *nip*." He played it again, and now I easily made out the phrase "nip it in the bud." (The speaker was the television character Barney Fife, on *The Andy Griffith Show*— although I didn't know that until he told me.) "When I gave you one word—one little bit of context—the whole thing snapped into place," he said. "You can do the same thing, in school, with kids who have trouble hearing, by giving them a vocabulary list in advance, or having them read a chapter before a lesson. It's like listening to music on AM radio. If you already know the song, you can actually enjoy it, even with a signal that's completely ripped apart."

All this is true for machines as well. Andy Aaron is a researcher at IBM. One of his projects in recent years has been helping to create the voice of Watson, the company's *Jeopardy!*-beating artificial-intelligence technology. "There are two systems at work in computer speech-recognition," Aaron told me. "First, there's the acoustic model, which listens to sounds and decides what phoneme it just heard. You say 'school' and it hears the four phonemes S-K-OO-L. But that isn't sufficient. It also needs a language model—because if you say 'M-AY-L,' how does it know whether you meant 'mail' or 'male'? So it looks at the surrounding words for context. If the word was preceded by 'deliver the . . . ,' then the computer knows you meant 'mail.' As you speak, the system checks every word against the preceding words. Another way to think of it is to say that system is constantly predicting what you're about to say. If you say 'I just got back from the . . . ,' the system has a list of words that will probably come next (*office, doctor, meeting*), and it has a list of words that are very unlikely to come next (*went, around, probably*)." There are even computer systems that analyze the movements of speakers' lips, to improve their recognition accuracy—and such systems exist because machines have the same problems that people do when they try to decipher sounds out of context. Aaron continued, "If you compose a sentence of random words, a speech-recognition program does a bad job of transcribing it, because it's no longer able to predict what's coming next."

All these factors affect comprehension. Many hearing people who have used Zoom, Skype, FaceTime, or similar online conferencing services to converse with family members, teachers, and business associates, during the coronavirus pandemic or otherwise, have observed that such sessions often seem more mentally and physically taxing than face-to-face interactions. One reason may be that online conversations lack many of the nonverbal contextual clues that enhance our understanding of what people say to us in person: with less clear information from facial expressions and other signals, we have to work harder to fully grasp what's being said. Hearing loss creates a similar impediment, in interactions of all kinds. One of the benefits of wearing hearing aids, for people who need them, is that it makes interacting with the world less physically taxing. Jason said, "There's a cognitive theory which basically says that our mental capacity is finite, like a glass of water, and we can allocate it among various activities—driving a car, having a conversation, reading a book. But if you have hearing loss you need a lot of that just for listening, and the more you need for listening, the less you have left for anything else you want to do."

LATER THAT DAY, Chris McCormick, Starkey's chief marketing officer, showed me a couple of hearing aids from the company's Halo line, the first version of which was introduced in 2014. Halo hearing aids have Bluetooth features that were developed in collaboration with Apple, and they can stream audio content from any Apple device. (Developing those features took Apple's engineers longer than they had thought it would, because they weren't used to working with such dinky voltages.) Halo hearing aids, like many other modern hearing aids, can also adjust automatically to different environments. They eliminate wind noise and reduce background sounds between spoken syllables during conversations in crowded places, and they can be used in conjunction with a smartphone app that makes it possible for them to do things like switch to a customized automobile mode as soon as the phone's accelerometer detects that the wearer is traveling in a car. McCormick said, "If you regularly visit a Starbucks, you can fine-tune a setting for that particular environment—the barista grinding coffee beans, other customers talking—and then geotag it, so that when you pull into the parking lot your hearing aids will switch to that mode." They also have a feature called Find My Hearing Aids, which uses a signal display on your phone to let you know when you're getting warm. Just the week before, a visitor had used that feature to find a hearing aid they'd lost on a Starkey shuttle bus.

McCormick also showed me hearing aids from Starkey's Sound-Lens Synergy line. Each unit looks scarcely larger than the battery it runs on—like a short, fat candy corn. The model he showed me is too small for Bluetooth, but it can be inserted deep into an ear canal. "A lot of people don't want hearing aids to be visible at all," he said. "There's still a vanity issue that people face, and that's why we make devices that are completely invisible. It's like a contact lens for your ear, if you will." He pushed one into his own ear and turned his head to the side; I could see no part of it, even from close up. I asked him how he could possibly remove it, and he showed me: by pulling on a snippet of nylon filament. "It's kind of like fishing line, with just a little ball at the end, and it's virtually unbreakable," he said. "You could pull your ear off before you could break that little line."

Since then, Starkey has introduced several more new products, among them hearing aids called Livio AI (as in artificial intelligence). The aids contain Fitbit-like inertial sensors, which, according to Starkey, not only count your steps but also potentially "monitor your body and brain health," by using a proprietary smartphone app to compute your "Thrive Wellness Score." Starkey says the aids can perform near-simultaneous translations of twenty-seven foreign languages—although that function requires a smartphone, an internet connection, and tolerance for the frequently comical limitations of Google Translate. The aids also permit users to adjust a few performance settings—again, with the help of a smartphone.

A DISCOURAGING FACT ABOUT HEARING AIDS is that people with hearing problems who do finally visit an audiologist, and agree to spend thousands of dollars, often feel exasperated from the outset. They wear their fancy aids for a few weeks, then put them in a drawer and never touch them again, or they lose them and don't replace them, or they wear them only once in a while, as President Reagan initially did. When that happens, it's not because they're upset that their aids don't count their steps or enable them to converse with their Parisian taxi driver. It's because they're disappointed in the way their aids handle the one task that truly interests them: helping them hear better.

My own aids do something annoying that I didn't notice until a day or two after I'd received them: in quiet environments they make a constant audible *SHHHH* sound, which jumps in volume slightly in response to sudden noises, like snapping my fingers, typing, clicking on a light, or turning the page of a book. The reason is that the aids react to silence by turning up the gain, and then react to noise by quickly, but not instantaneously, turning it down—resulting in a steady background whoosh that jumps in volume slightly, after a brief delay, in response to discrete sounds. I notice it only in relative silence, and if my hearing were worse I might not notice it at all. But it bugs me enough to make me not want to wear my hearing aids in situations where they might help.

The most common source of disappointment with hearing aids is that even the most expensive ones don't correct faulty hearing in the same way that eyeglasses correct faulty eyesight. If your vision is blurred because your eyeball is shaped in a way that prevents light from landing correctly on the retina—that is, if you have myopia, hyperopia, astigmatism, or presbyopia—being fitted with the right corrective lenses (or undergoing a surgical procedure like Lasik) can give you perfect vision, enabling you to see the way people who don't need glasses see. When my wife got her first pair of glasses, in second grade, she shouted, with astonishment, "I can see inside that truck!"—and a friend, equally astonished, asked, "Did they give you X-ray vision?" I had a similar experience, in fifth grade, when I realized that my brand-new glasses enabled me to make out individual leaves on trees. Hearing aids don't do the equivalent with sound. If you've lost all ability to detect frequencies above 5,000 hertz, no hearing aid can give that back. A hearing aid can turn up sounds you now detect only faintly, but they can't transform you into the person you were before you discovered rock and roll.

Even technological breakthroughs can constitute an impediment to hearing-aid use. Modern batteries are remarkably small—an amazing achievement unless you have arthritic fingers or an age-related tremor, in which case prying open your hearing aid's plastic battery door, removing the old battery, extracting the new battery from its dispenser, peeling away the protective adhesive tab, and snapping everything closed again can be as difficult as threading a needle in the dark. For decades, the major manufacturers have focused on making hearing aids smaller and less conspicuous, and on adding features that are only tangentially related to hearing loss. One result has been significant increases in price—often more than $3,000 per ear—without corresponding increases in satisfaction.

AT THE TIME OF MY VISIT TO STARKEY, the company was in the news for reasons unrelated to miniaturized hearing technology. In September 2015, William Austin, the founder and chairman, fired several employees, among them Jerry Ruzicka, the president, and Scott Nelson, the chief financial officer. Two months later, federal agents raided Ruzicka's and Nelson's homes, and in 2016 the Department of Justice indicted the two of them and three others for what the U.S. attorney called "a massive and long running fraud scheme against a corporation." According to the Justice Department, the defendants, for a decade, had "conspired to embezzle and misappropriate money and business opportunities belonging to Starkey and Sonion, a major supplier of hearing aid components to Starkey."

The trial began in January 2018, and during it Austin didn't do his own reputation any good. He testified that he was only minimally involved in the operation of the company—not a secret at Starkey—and, among other interesting revelations, that he had once spoken to an angel. The trial lasted six weeks, and shortly before it ended, the judge struck some of Austin's testimony, on the grounds that it had been shown to be false. Nevertheless, Ruzicka was convicted of eight counts of fraud and was later sentenced to seven years in prison, and one of the other defendants was convicted of three counts and was sentenced to two years. Nelson, before the trial began, had pleaded guilty to conspiracy to commit fraud and had testified against the others; he was sentenced to two years as well.

In 2017, Austin chose Brandon Sawalich, the son of his fourth wife, to succeed Ruzicka as president, despite the fact that many at the company viewed him as unqualified for that job. Michela Tindera, in an article in *Forbes* in 2018, wrote that Scott Nelson, during the trial, had "testified that Sawalich got the company to pay for a variety of personal expenses up until about 2011—submitting bills for an ice skating rink, a chicken coop, lawn services, dog boarding and even a fish-tank cleaning." Nelson said, furthermore, that Sawalich had a reputation within the company of being a "serial harasser" of women, and Tindera cited accounts of several incidents, including an alleged sexual assault of a young employee in 2001, which resulted in a lawsuit that was settled in 2003.

None of this has been good for the company. Starkey's sales are believed to have declined since its troubles were made public, its share of VA hearing-aid purchases has fallen, and it faces continuing legal problems, including the possibility of a lawsuit by employees. More than anything else, the Starkey trial revealed a lot about the economics of the hearing-aid business, which was shown to be profitable enough not only to have made Austin a billionaire but also to have generated so many excess millions that company insiders were tempted to grab them. The other major hearing-aid manufacturers may have seen Starkey's troubles as creating a competitive advantage for themselves, but in the long run they will suffer, too, if potential customers conclude that, no matter whose products they decide to buy, they're being ripped off.

Eight

STIGMA

In 2013, the former television personality Charlie Rose devoted an entire PBS program to hearing loss, and during the broadcast two of the participants—Eric Kandel, a scientist who won a Nobel Prize in 2000, and Rose himself—were wearing hearing aids. Yet neither Kandel nor Rose mentioned that fact, even though the program lasted nearly an hour and hearing aids were a major topic of discussion. David Corey, the Harvard Medical School professor who showed me what hair cells look like, appeared on the program, too, and got a good look into Rose's ears; I myself could see Kandel's hearing aids, on the screen. Kandel is in his late eighties. Does he believe that people would think less of him and his Nobel Prize if they knew that he can't hear as well as he did when he was younger? And Rose has spoken openly about his heart problems. Why wouldn't he mention his own direct experience with the topic of his show?

Robert Dobie, the San Antonio professor I spoke with, told me, "It's possible nowadays to have hearing aids that are pretty unobtrusive, but that stigma is still a factor, and it's greater among men than among women. Hearing aids, for some people, signal decrepitude, loss of vigor, loss of competence. People wear glasses, in many cases, from adolescence onward, or even before, and while glasses may not always be thought of as attractive, they are not thought of as a sign of diminished competence. Hearing aids, for most people, really are a sign of age, rightly or wrongly." Children who wear glasses are occasionally picked on for that reason ("Hey, Four-eyes!"), but on television and in movies glasses are often a signifier of intelligence—and not only in nerds, mad scientists, and absentminded professors. An easy way to suggest that you are more than a supermodel or a Hollywood star or a professional athlete is to wear glasses when you talk to reporters. People with perfect vision sometimes wear glasses because they like the way glasses look.

The hearing-aid stigma is much older than hearing aids. It arises from the nature of deafness: the inability to hear is a serious impediment to the acquisition of language, and language is what we think of as the thing that distinguishes us from beasts. The assumption for centuries was that people who couldn't speak (because they couldn't hear) must also be unable to think, and were therefore only nominally human: they were "dumb." And that prejudice was, and is, self-reinforcing, because successfully overcoming the obstacles posed by hearing loss, even today, requires much more than doing the equivalent of putting on a pair of glasses.

ALICE COGSWELL WAS BORN IN HARTFORD, Connecticut, in 1805. Two years later, she lost almost all her hearing to cerebrospinal meningitis, which, in addition to being a significant cause of deafness, was common enough to be known by multiple names: spotted fever, typhus syncopalis, peripneumonia notha, catarrhal fever, winter epidemic. Alice's father, Mason Fitch Cogswell, was a distinguished physician—he performed the first successful cataract-removal operation in the United States—but there was nothing medical that he could do to help his daughter. In 1812, he asked Connecticut's General Association of Congregational Ministers to conduct a census of deaf people in the state. They did, and counted eighty-four, and he used their findings to campaign for the creation of a school. This was not a universal concern. Deaf children were almost always viewed as hopelessly impaired, and even in well-to-do families little consistent effort was made to educate them. Cogswell was shrewd to seek the help of a religious organization, rather than a medical or educational one, since most of the early pressure for deaf education came from Christian clergy, who worried that people who could neither hear sermons nor read Scripture had no hope of salvation. John of Beverley, a Catholic saint, was canonized in 1037 in part because he had taught a deaf boy to speak a few words—an accomplishment that for centuries was viewed as literally miraculous.

In the spring of 1814, when Alice was nine, Thomas Gallaudet, a son of the Cogswells' next-door neighbors, visited his parents while recovering from an illness. He had earned two degrees at Yale and was studying for the ministry at Andover Theological Seminary. At some point, he noticed that Alice wasn't playing with other children, and was told of her condition. He handed her his hat and used a stick to make the letters H-A-T in the dirt. She grasped the concept quickly, and, just as quickly, Gallaudet decided that his future lay not in a pulpit but in a school for the deaf.

That, at any rate, is the official version of the story. The narrative has surely been condensed and semi-mythologized, but it must be reasonably accurate, because Alice's father and a group of prominent friends put up money to send Gallaudet to Europe, where schools for the deaf had existed for decades. (Samuel Johnson, in his account of the trip that he and James Boswell took through Scotland in 1773, wrote, "There is one subject of philosophical curiosity to be found in Edinburgh, which no other city has to shew; a college of the deaf and dumb, who are taught to speak, to read, to write, and to practise arithmetick.") Gallaudet's assignment was to study the European schools, and to come back and establish something similar in the United States.

He traveled first to England, where a family called Braidwood operated what was essentially a for-profit monopoly in deaf education. The Braidwoods told Gallaudet that they would teach him their methods, but that when he returned to America he would have to keep those methods secret and pay the Braidwoods a royalty for each student he taught. While he considered that proposal, he attended a lecture, in London, by a Catholic clergyman who ran a school for the deaf in France. Gallaudet later visited that school and, in 1816, persuaded one of its students, Laurent Clerc, to return with him to the United States. The following year, in Hartford, Gallaudet and Clerc opened the Connecticut Asylum for the Education and Instruction of Deaf and Dumb Persons, in a rented room in a hotel on the site of what's now the corporate headquarters of an insurance company. The inaugural class consisted of seven students of various ages, among them Alice Cogswell, who was then twelve years old. Clerc was the teacher.

The school's name changed over the years, and actually got slightly longer—American Asylum, at Hartford, for the Education and Instruction of the Deaf and Dumb—before becoming significantly shorter. It is now simply the American School for the Deaf. Since 1922, its campus has been in West Hartford, a prosperous residential suburb, on fifty-four acres of what used to be farmland, a few miles from the original site. It has 144 students between the ages of three and twenty-one, and, of those, ninety-two are residential. ASD was the "mother school" of all deaf schools in the United States—including Gallaudet University, in Washington, D.C., whose first president was the youngest of Thomas Gallaudet's eight children.

THAT THOMAS GALLAUDET ENDED UP working with the French rather than the English is historically and pedagogically significant. The Braidwoods were primarily what are known now as "oralists": they taught their deaf students to read lips and to speak. That technique worked, to some extent, with children who had residual hearing, but it was ineffective with children who were prelingually deaf, meaning that they had become deaf before they acquired language, or who were profoundly deaf for other reasons. The French, by contrast, were "manualists": they communicated primarily through sign language. During Gallaudet and Clerc's return trip across the Atlantic, which took six and a half weeks, Gallaudet taught Clerc some English, and Clerc gave Gallaudet lessons in French sign language.

The sign language that Clerc used was based on one that had arisen organically, over multiple generations, among deaf people in Paris. The head of the French deaf school had realized that successful teaching depended upon fluent communication, and that, for those who couldn't hear, signing was far more effective than speaking and lipreading. But he had encumbered the existing system with his own "methodical" modifications, the purpose of which was to enable users to exactly transcribe written and spoken French, including all the complexities of French word order, gender, and syntax. (The cases of nouns, for example, were indicated by rolling one index finger around the other in particular ways.) Gradually, though, Clerc, his students, and other teachers at ASD abandoned those impediments and incorporated signs and signing concepts that the school's students had brought with them from other places. What's now known as American Sign Language—the principal manual language of the deaf in the United States and in the English-speaking parts of Canada— began in those classrooms at ASD, and grew through the evolutionary process that linguists call "language contact"; it was a joint creation of the school's students and teachers.

Some of the most important contributions to American Sign Language were made by students from Chilmark, Massachusetts, on the island of Martha's Vineyard, off Cape Cod, who made up a significant fraction of the school's early enrollment. Chilmark at that time had what was probably the highest incidence of congenital deafness of any place in the United States. By the late 1800s, something like 4 percent of the town's residents (and a quarter of the residents of the Chilmark village of Squibnocket) were deaf. The reason wasn't understood at the time, but we know now that the concentration was caused by a specific recessive genetic mutation, whose effects had been multiplied, over generations, by the limited marital opportunities available within what was then an isolated farming and fishing community.

The underlying mutation is believed to have originated in a similarly isolated agricultural community in England, near Kent, and to have been brought to the Vineyard by more than one early settler.

I've spent part of every summer in Chilmark for a little over forty years, and my wife has done the same for fifteen years longer than that, but even by the time she first visited, in the early 1960s, Chilmark's deaf community no longer existed. The fact that it had disappeared is partly a tribute to the power of effective education: many non-hearing Chilmark residents had attended ASD and then found jobs off-island, some of them as teachers of the deaf—and even those who returned to the island increasingly married outside of the genetic enclave they'd been born into, as Martha's Vineyard became more populous and more diverse. The last Chilmark resident whose deafness was traceable to the original genetic mutation died in 1952.

I knew nothing about Chilmark's history with deafness until I came across an undated newspaper clipping, apparently from the early 1920s, in a crumbling scrapbook on a shelf in an old barn at the place my wife and I go every summer. The article appeared to be a reprint of an item originally published in the *Boston Herald*. The writer, Ethel Armes, described Chilmark as a town in which the postmaster, storekeeper, and pastor were all deaf, and recounted the experience of a new resident who needed potatoes: "We carried the last remaining one to the farm next to us to show what we wanted. The farmer's wife went into the house, got a telescope, and signaled to another farmhouse further up the hillside. Very soon that neighbor appeared, also armed with a telescope. After some brief signaling we had a bushel of potatoes at our door. In every Chilmark family there is a telescope—and also there is a retired sea captain." The women were signing to each other, and their sign language, in combination with their telescopes, had given them the equivalent of telephone service at a time when telephones were a rarity on the island.

During a visit to Martha's Vineyard in the late 1970s, Nora Ellen Groce, a graduate student in anthropology at Brown, was given a tour by an old-timer, who told her stories about deaf residents of Chilmark many years before, and she realized that the deafness he described must have been hereditary. She spent several years conducting interviews and doing research, in both the United States and England, and turned her findings into her doctoral dissertation, which Harvard University Press published, in 1985, as *Everyone Here Spoke Sign Language*. (She's now a professor of epidemiology and public health at University College London.) Her book is excellent—both as an anthropological detective story and as an inspirational account of human beings acting like human beings. In Chilmark for most of three centuries, Groce concludes, the inability to hear carried no stigma. By the early years of the eighteenth century, at least, nearly everyone in the area learned to sign as a matter of course, beginning in early childhood, and almost all hearing residents could move back and forth between English and signing. (Groce hypothesizes that the sign language they used was based on one that had originated in the part of England from which the island's first deaf settlers had migrated.) Indeed, signing in Chilmark was so unremarkable that older residents, when Groce interviewed them years later, sometimes had trouble recalling who had been deaf and who hadn't. The inability to hear seemed unexceptional, and people born on the island who later traveled to other places were often surprised to find that deafness was far less common elsewhere than it was at home.

Groce quotes at length from an essay, written in 1861, by a sixteen-year-old Chilmark resident, about a picnic she'd attended when she was seven or eight: "We put our sweet cakes on the long table and there were many kinds of cakes, pies, oranges, cherries, lemonade, and beautiful flowers in glasses on it. I played with some girls and boys on the hill for pleasure. Some of the children told me about the clams in the ground and we ran to a place where clams were baked for the people to eat. . . . When we had done we all walked pleasantly to the sea to look at it for a little while or talked to each other and we had an excellent Picnic." Nothing in the girl's essay hints at what is surely the most interesting fact about her: that she'd been deaf since birth. (The essay was a writing assignment at ASD, where the girl had been a student for four and a half years.) Groce found that every Chilmark family had a direct, multigenerational connection to deafness, and that the people she interviewed, in their recollections, often made no distinction between speaking and signing. She quotes a resident's description of an argument with an elderly neighbor: "She yelled at me, and I told her off, but good! Come to think of it, I guess we did our yelling in sign language." Hearing residents sometimes signed to one another even when no deaf person was present—when telling the punch line of a dirty joke, or when communicating with a fisherman on another boat (arms extended overhead, to make the hands more visible at a distance), or when talking behind a teacher's back in school, or when chatting with a sibling or a friend across the room at a town meeting. A visiting preacher remarked, after a church service, that a woman sitting in the front pew had nervously fidgeted with her hands throughout his sermon, but was told that she'd merely been interpreting for her husband, who was deaf and was sitting next to her. (The woman, when in church, signed inconspicuously, in her lap—"just about the same as a person would if they were knitting a sock.")

Perhaps the most remarkable fact is that in Chilmark there were no activities from which the deaf were excluded, and no activities that were conducted exclusively for the deaf. "On the mainland profound deafness is regarded as a true handicap, but I suggest that a handicap is defined by the community in which it appears," Groce concludes. "Although we can categorize the deaf Vineyarders as disabled, they certainly were not considered to be handicapped." She describes a deaf farmer who, once automobiles began to appear on the island, was able to tell when a vehicle was approaching his wagon from behind by watching for particular movements of the ears of his horse. "They participated freely in all aspects of life in this Yankee community," she continues. "They grew up, married, raised their families, and earned their livings in just the same manner as did their hearing relatives, friends, and neighbors. As one older man on the Island remarked, 'I didn't think about the deaf any more than you'd think about anybody with a different voice.'"

THE GOLDEN AGE OF SIGN LANGUAGE officially ended six decades after ASD's founding, at the Second International Congress on the Education of the Deaf, held in Milan in 1880. Delegates there endorsed a concept that had been gaining adherents in the United States and elsewhere—that of "the incontestable superiority of articulation over signs in restoring the deaf-mute to society and teaching him a fuller knowledge of language"—and they voted to ban the use of signed languages in all schools for the deaf, worldwide. Douglas C. Baynton, in *Forbidden Signs: American Culture and the Campaign Against Sign Language*, published in 1996, characterizes the anti-signers as "a generation frightened by growing cultural and linguistic diversity," who "thought in terms of scientific naturalism, especially evolutionary theory," and "associated sign language . . . with 'inferior races' and 'lower animals.'" ASD and Gallaudet University both resisted the oralism-only movement, in ASD's case by continuing to teach signing alongside other methods. But the character of deaf education changed radically, and the stigmatization of the deaf became more intense and more devastating at a moment when history could easily have turned in a different direction.

Oralism obviously didn't entirely replace manualism, since ASL still exists, and even flourishes, today. But signing of any kind by the deaf came under tremendous stress for most of the following century. One consequence was that, at most deaf schools, deaf teachers were replaced by hearing teachers, and what had once been communities of the deaf were broken up. One of the leaders of the anti-signing movement was Alexander Graham Bell, whose mother was deaf and whose best-known invention, the telephone, he conceived of while thinking about potential technological aids for the hard of hearing. Bell argued that teaching the deaf to sign and educating them in residential schools encouraged them to marry one another, perpetuating their disability and thereby weakening the human species. Bell—like Alfred Binet, the inventor of the IQ test, and Carl Brigham, the inventor of the SAT—was a eugenicist. In 1883, in a paper he presented at a meeting of the National Academy of Sciences, in New Haven, he argued: "If the laws of heredity that are known to hold in the case of animals also apply to man, the intermarriage of congenital deaf-mutes through a number of successive generations should result in the formation of a deaf variety of the human race." He worried less about certain other conditions that he also believed to be hereditary, because "we do not find epileptics marrying epileptics, or consumptives knowingly marrying consumptives." The deaf, by contrast, posed what he believed to be a unique threat to the vigor of the human gene pool—and that threat was magnified by the fact that the deaf had a method of communication that the hearing could not understand (residents of Chilmark excepted).

There are humanitarian arguments for oralism, the goal of which is to integrate the deaf into a world in which many more people speak than sign. But historically those arguments have been made far more often by the hearing than by the deaf. And because the arguments emphasize speech, rather than communication, they have often had the effect of isolating the deaf, not only from the hearing but also from one another—as Bell and others overtly intended. At the congress in Milan, the fundamental insight of the French school that Gallaudet visited and Clerc attended—that education can't occur among people who don't understand one another—was essentially abandoned.

A resurgence of respect for manualism began in the 1960s, when William Stokoe, a (hearing) linguist and the head of the English department at Gallaudet University, argued persuasively, in two landmark books, that ASL and other sign languages are real languages. Hearing people tend to think of signing as a form of pantomime, or as a set of physical ciphers—like the signals that the caddies at one of my favorite golf clubs use, from far up the fairway, to tell golfers where their tee shots have just ended up ("tall rough," "bunker," "water hazard," "fucked behind a tree"). But ASL and other true sign languages have grammar, abstract symbols, and a complex underlying structure, just as any spoken language does. In many ways, sign languages are more versatile and expressive than spoken languages, because the possibilities for nuance and individual variation are greater. English syntax is linear: word follows word follows word. ASL is sometimes described as four-dimensional, since it simultaneously incorporates facial expressions, body positions, and any number of spatial and temporal relationships among the fingers, hands, arms, and other body parts. A downside is that ASL is tough to record, and therefore to preserve, except on video. (No one today knows the sign language that was used on Martha's Vineyard, even though it survived into the twentieth century.) The signing equivalent of the printing press is probably YouTube.

Proof that sign languages really are languages comes from neurology. As Oliver Sacks writes in *Seeing Voices: A Journey into the World of the Deaf*, published in 1989—still a fascinating book, even though it contains more footnotes than *Infinite Jest*—deaf people who suffer damage to the brain's speech centers, in the left hemisphere, retain the ability to make "the non-grammatical expressive movements we all make (shrugging the shoulders, waving goodbye, brandishing a fist, etc.)," which are controlled by the right hemisphere, but lose the ability to sign. In other words, they're still able to make manual signals like the ones used by the caddies I mentioned, but not to use ASL. Sign language isn't a limitation or a "crutch"; preventing people who need it from learning it and using it has never been good for anyone, and certainly not for "society."

During the past couple of decades, ASL and deaf education itself have come under renewed pressure—and, paradoxically, this new pressure has come primarily from improvements in technology intended to help the hard of hearing, including hearing aids and cochlear implants. Most people think of innovation as a good thing, but, as is often the case with technology, and also with human beings, the subject is complicated, and the most important consequences often turn out to be unintended ones. I'll have more to say about all of that in chapter eleven.

Nine

BEYOND CONVENTIONAL
HEARING AIDS

Way back in the early 1980s, when I was a senior writer at *Harper's*, I wrote an article about what you ought to do with your body once you've finished with it. Among many other things, I learned why dying is so expensive. The funeral industry has always invested heavily in lobbyists, and, as a consequence, many states don't allow you to dispose of your corpse without the involvement of a funeral director, and most cemeteries require that coffins be enclosed in concrete grave liners or vaults—in effect, coffins for coffins. Bodies that are going to be cremated often have to have coffins, too, and there are innumerable other requirements and customs and practices and procedures that seem to exist only to run up the tab. When I went funeral shopping for a fictitious ailing aunt, the salesman urged me to test the firmness of the (name-brand) mattress inside a coffin that he said was considerably less "minimum" than the one I'd expressed an interest in.

The hearing-aid business has also been structured to maximize profits and reduce competition. The FDA classifies a hearing aid as a medical device, which it defines as "an instrument, apparatus, implement, machine, contrivance, implant, in vitro reagent, or other similar or related article, including any component, part, or accessory," and so forth, for another eighty-one words. The FDA imposes requirements on how hearing aids can be dispensed, and states add requirements on their own. For decades, you haven't been able to buy a hearing aid without first being evaluated by a medical professional (something that isn't required when you buy other audio-enhancement devices, such as radios, telephones, and television sets). The services of audiologists are built into the price, and if you're dissatisfied with almost anything about the way your hearing aid functions, you can't make more than minor adjustments by yourself, even on super-expensive recent models.

That began to change in 2017, when Congress passed a bill requiring the FDA to establish a category of hearing aids that consumers would be able to buy over the counter and calibrate on their own. As of early 2019, the FDA had approved at least one such hearing aid (about which more below), and by the time you read this you will probably be able to buy it, or something like it, for considerably less than traditional hearing aids cost now. A revolution in the selling and fitting of hearing-improvement devices is now under way. As the scientist I quoted at the end of chapter one told me, "There is no better time in all of human history to be a person with hearing loss."

Dianne Van Tasell—the University of Minnesota professor who persuaded Jerry Ruzicka to hire her to form a team to develop a digital hearing aid at Starkey—left the company in 2002. "By then I had seen how the sausage is made," she told me. "The thing that really bothered me was the humongous profit margins that the whole system had racked up, to the detriment of consumers. The companies had worked with audiologists to create and maintain state licensure laws, which drove consumer price so high that very few people could afford them." She was especially annoyed that no one in the hearing-aid industry was interested in allowing consumers to adjust their own devices, beyond making minor tweaks. "Hearing aids are not rocket science," she continued. "We're not implanting them in your head or sticking electrodes in your ear. If you can adjust a graphic equalizer on a stereo to make your music sound good, you can adjust a hearing aid." But executives at Starkey and other companies told her that they were not interested in any innovation that would undermine their traditional business model. "That's when I said, O.K., I don't want to be part of this; I want to be part of something that actually makes progress."

After leaving Starkey, Van Tasell worked with several start-up companies that hoped to produce hearing aids that would be both self-adjustable and as easy to shop for as telephones or fitness trackers. She told me, "Without exception, those companies had great ideas and great scientists, but they didn't understand regulation, and they had failed to see the extent to which they would be crushed by the current system because everyone in it is so fat and happy." So she retired, she and her husband moved to Tucson, and she signed up for courses (in the web-page language HTML) at a community college.

One day, though, she got a call from someone at IntriCon, a small Minnesota company that describes itself, on its website, as a creator of "miniature and micro-miniature body-worn medical and electronics products." IntriCon had been hired by United Healthcare, a major insurance company, which had decided to provide hearing aids, as a benefit, directly to the people it insured, without the involvement of audiologists. IntriCon had agreed to manufacture the devices, but before United Healthcare could distribute them it needed to be able to fit the recipients over the internet. IntriCon hired Van Tasell to devise an online hearing-evaluation method, and in less than a year she and the team she assembled had done that.

"During this time, I told the people at United Healthcare that the FDA was going to be all over them," she told me. "And they said, 'Don't worry—we have all these lawyers.' So we developed our method, and they launched their program, and immediately there was a huge outcry from the hearing industry, and within about four weeks the FDA had issued a cease-and-desist order. It all happened because United Healthcare had made the mistake of thinking their in-house lawyers knew something about regulatory issues. But they didn't, and they got nailed."

United Healthcare's hearing-aid program, though terminated by the FDA, convinced Van Tasell that self-fitting was a viable idea. She described what she'd done at IntriCon to Sumit Dhar, who is the chairman of the Department of Communication Sciences and Disorders at Northwestern, and has long been interested in understanding why so few people who need help with their hearing actually seek it. Van Tasell suggested that they apply together to the National Institutes of Health for a consortium grant to study the problem, and Dhar introduced her to Andy Sabin, a graduate student who had been thinking about ways to provide user controls to people with hearing aids. "Andy whipped out his smartphone and said, 'Oh, yeah, I've already done something like that,'" Van Tasell told me. "I looked at it and said, 'Oh my god, yes!' I congratulated him, and told him that what he had done was great, and said that now I could go back to my community college—but he said, 'Wait, wait! I don't know how to do anything with it!'"

LIKE MANY OTHER PEOPLE who work in acoustics, Andy Sabin arrived at his calling through music. "I'm sort of a failed musician, or a mediocre musician, and probably as a consequence I was more attracted to the technology of music-making," he told me. "Even in high school, I had a recording studio in my basement, and throughout college I did recording on the side, and I studied the auditory system." As he worked with musicians, he was struck repeatedly by different versions of what turned out to be a common issue: "I'd be sitting in the recording-studio booth with, say, a band's guitarist, and they would say, 'You've got to make my guitar sound more 'cloudy,' or some really high-level adjective like that." At first, Sabin dismissed such requests as proof that the musicians didn't know what they were talking about—"There's no 'cloudy' dial on a mixing board"—but eventually he decided that they were using imprecise language only because they didn't have enough technical knowledge to specify exactly which acoustic parameters would have to be adjusted in order to re-create the sound they could hear in their head. "And I found that to be a really interesting problem," Sabin continued. "All the tools we have to make adjustments like that are focused on engineering. There are parameters that are tied to signal processing and bandwidth and frequency regions—but that's not how artists think."

Sabin was working on his dissertation, and as a side project he quizzed musicians about sounds they wanted to create, then played multiple examples. "I would ask them, 'Hey, is this "cloudy"? Is that "cloudy"? How "cloudy" is this?' And then I would just sort of let the computer figure out the relationship between the signal-processing parameters and the sound they were looking for. Once I'd done that, I could give them a 'cloudiness' dial." He created a plug-in for Pro Tools and GarageBand, which are programs that function like music-studio mixing boards, and he did something similar that worked with microphones. Neither project amounted to much, in terms of sales, but he realized that the software he'd created could be modified to work with hearing aids.

"The problem for people who have trouble hearing is exactly the same as the problem for musicians," Sabin told me. "One reason hearing aids are so expensive is the required involvement of audiologists. In effect, the high cost of the hearing aid subsidizes the time of the audiologists, whose role in hearing-aid fitting is extremely similar to the role of the recording engineer in music production." Sabin used the example of someone who wears hearing aids in a noisy restaurant and doesn't like the way they sound in that environment. "So they call their audiologist and schedule an appointment, for two weeks later. Then, at the appointment, they say, 'Hey, my hearing aids sounded bad when I was in a restaurant two weeks ago.' And the audiologist says, 'Well, what do you mean by "bad"?' And they go back and forth about that, and then the audiologist makes an adjustment based on their conversation, and says, 'Would this have corrected the problem you had two weeks ago?'" This difficulty is made worse by the fact that people who wear conventional hearing aids don't always realize that the high price they paid their audiologist almost always includes follow-up visits and other continuing services—benefits that audiologists have no financial incentive to explain.

To both Sabin and Van Tasell, this seemed absurd. Sabin said, "Wouldn't it be easier to have an app or a remote control or a button that would let you adjust a hearing aid on the spot—to make it sound good right then? It seems obvious, right?" James Gold, my investment-banker friend who has both hearing loss and severe tinnitus, makes his audiologist adjust his hearing aids in actual restaurants. He reserves a table for five, and the audiologist sets up his equipment, and they fiddle with settings over an expensive dinner. But that option isn't available to everyone—and dealing with truly noisy environments isn't something that hearing aids are necessarily great at anyway.

The Centers for Disease Control and Prevention has a vast database of hearing-test data, which it has collected over many years. Sabin used a computer to analyze the audiograms in the database, and discovered that hearing losses are not randomly distributed across the full range of humanly detectable frequencies. The same patterns recur, and most of the variability is statistically narrow. "Hearing loss has stereotypical shapes—so stereotypical that there are actually ISO standards, from the International Organization for Standardization, for things like '65-year-old male' and '70-year-old female,'" he said. "As a result, even though there is a huge, huge number of possible combinations of hearing-aid settings, the range of settings that can actually help someone is very, very small."

This predictability makes the adjusting of hearing aids by audiologists more automatic than most hearing-aid users suspect. The rise of digital hearing aids made it possible to selectively add or remove amplification in specific narrow frequency ranges, but because the patterns are so predictable, audiologists don't always do that. Instead, they rely heavily on standard, research-based programs that dictate complete setting profiles determined solely by patients' audiograms. Two such programs are used almost universally: one that was developed in Canada, and one that was developed in Australia.

"Usually, manufacturers will make one or the other of those programs available to a dispenser—to the audiologist using their fitting software," an audiologist told me. "And sometimes a manufacturer will make up its own program. So the audiologist enters your audiogram into the fitting software, and sticks your hearing aid in there. Then they choose the Canadian program or the Australian program or the proprietary program, and all the settings are loaded automatically. That's how you fit a hearing aid." Bluetooth-enabled hearing aids, like the ones that Starkey developed with Apple, allow users to make some adjustments on their own, through proprietary apps on their phones, but the adjustments that users have access to are limited. The main work of fitting, for the vast majority of hearing aids, is done automatically, by whichever program the audiologist uses. Ideally, the audiologist also adjusts and validates the programmed settings by inserting a slender tube connected to a microphone all the way into the ear canal, beyond the hearing aid, and playing sounds from an external speaker. The audiologist then adjusts the programmed volume levels based on how loud the hearing aid is actually making those sounds next to the eardrum, to make sure the hearing aid is delivering as much gain as it's supposed to across the full range of frequencies. The reason for taking such "real-ear measures" is that the size and shape of an ear canal can affect how a hearing aid performs: the output from two identically programmed devices will sound louder in a small ear canal than in a large one. In addition, suddenly hearing diminished frequencies at a higher volume can be unnerving for people who haven't heard those sounds in a long time. Their brains will adapt in time, but if the proper amount of gain isn't added they'll never hear as well as they might have. Such adjustments can make the difference, for the recipient, between sticking with new hearing aids and throwing them into a drawer. (Even so, audiologists don't always take the time. When my mother-in-law was having trouble with her hearing aids, I suggested that she ask her audiologist to do a real-ear test. She did, and he told her that the test was complicated and that there was no need to do it.)

As Sabin and Van Tasell worked with the CDC's hearing data, they realized that it would be possible to reproduce almost everything an audiologist does, by creating a user-operated app that had just two virtual dials—analogous to the app that Sabin had created for musicians. They called one of the dials "loudness" and the other "fine-tuning." By turning the "loudness" dial, users were doing more than adjusting volume; they were actually scrolling through the equivalent of a full range of preprogrammed settings, like the ones in the standard Canadian and Australian programs. But Sabin and Van Tasell didn't believe that users needed to know that in order to adjust their devices. "The idea was to make the controller so simple that people would merely turn one dial until they could hear pretty well, and then tweak it with the other, and then they're done," Van Tasell said. They called their controller EarMachine.

WHILE SABIN AND VAN TASELL were working on this project, the National Institutes of Health serendipitously requested proposals for Small Business Innovation Research grants aimed at increasing the affordability and accessibility of hearing aids. They applied successfully, and, at the same time, realized that if they were going to turn EarMachine into a company, they needed a partner with a background in business. Van Tasell, coincidentally, had recently heard from Kevin Franck, an audiologist she'd met while he was working on his PhD. Franck had an MBA besides his PhD—an unusual combination for a hearing scientist—and he was working for a strategy consulting firm that provided marketing advice to start-ups. Even better, he had been involved in a project whose aim was to enable people with cochlear implants to adjust their own devices. "It was a perfect fit," Sabin said.

With their grant, the three conducted controlled tests of large groups of hearing-aid users, and those studies confirmed their belief that the users could adjust their own devices at least as well as audiologists could—something that Franck had also found to be true of people with cochlear implants. Franck told me, "We took a bunch of people and had an audiologist fit them, and then gave them our app and had them fit themselves, in a double-blind study. And we found that their settings were no different from the audiologists' and, in fact, that they preferred their own settings." In addition, Franck, Sabin, and Van Tasell concluded that people who adjust their own devices are less likely to give up on them and throw them into a drawer, because they feel both more in control and more invested.

Then Franck, Sabin, and Van Tasell began to approach

manufacturers. "Hearing aids are sort of universal," Van Tasell told me. "They're all the same under the hood. We told the manufacturers that if they would just open their hood to us, we would stick our tools in there and provide this simple interface that would allow users to adjust all their own parameters. We told them that, after an audiologist did the initial fit, the users would be able to handle their own adjustments and wouldn't have to come back."

They thought that hearing-aid manufacturers would be eager to be involved, but they weren't. Franck called on all the Big Six companies, and the pattern was always the same. "Our first contact would be with the engineering team, and the engineers would seem excited about it," Sabin said. But then they would pass us up to the business people, and inevitably the business people would say, essentially, 'We like this, but we don't want to anger our customers.' And it took me a while to realize that when they talked about their customers they weren't talking about people who wear hearing aids. They were talking about audiologists." Officials at more than one company told them that they might be interested in buying EarMachine, but only to kill it. "They were completely unapologetic about that," Van Tasell said. "They'd say, 'Oh, yeah, what you've done would work, but audiologists wouldn't like it, so, no, we'd never do it.' We would ask them if they didn't care about market penetration, and they would say, frankly, that they did not, because they were making enough money as it was."

The three then turned to consumer-electronics companies—an experience that Van Tasell, in particular, enjoyed. "I was sixty-four or sixty-five at the time, and Andy and Kevin were much younger," she said. "We'd meet with guys in their thirties with little glasses and skinny pants—and it was always guys, by the way, except for the women who were serving coffee. So we'd go to a conference room, and these guys would have their laptops out, and it would be like Andy and Kevin had brought their mother. As a result, no one could be mean to me—and what was even funnier was that I was the one giving the science part of our presentation."

A company that all three partners were eager to work with was Bose, the audio-equipment manufacturer. It had been founded in the 1960s by Amar Bose, who taught at MIT for nearly five decades. Bose's father had fled to the United States from India in 1920, to escape what he believed would be his certain execution by the British. (He'd fought for Indian independence while studying at the University of Calcutta, and he emigrated just before he would have graduated.) Amar was born in Philadelphia in 1929. When he was young, he taught himself the rudiments of electrical engineering by salvaging and restoring broken train sets, and during the Second World War he helped to support his family by repairing radios in a workshop he'd set up in his parents' basement. (His father's business, importing Indian textiles, had been devastated by the war.) "I had a little pact with my father that if my grades remained good, I could go to school only four days a week, and he would write an excuse saying I had a headache or something," Bose himself recalled later. "The teachers all knew this; it was always on a Friday and so on Monday they'd ask me, 'How many radios did you fix, Bose?'"

He earned undergraduate and graduate degrees at MIT, in electrical engineering, then worked in the Netherlands for a year and spent the next year in India, on a Fulbright. He joined MIT's faculty when that fellowship ended. He said later that when he'd arrived at the school as an incoming freshman—after being rescued from the waiting list by an alumnus who told the admissions office about his radio repair shop—he found that he was far less well prepared, academically, than his classmates, and concluded that in order to survive he would have to work harder than anyone else. So he permitted himself just a two-hour break each week, on Sunday, during which he listened to the radio. His struggles as an undergraduate shaped his ideas about teaching and made him legendarily sympathetic to students who were slow to grasp difficult concepts. Toward the end of his MIT career—he retired in 2001—his courses were so celebrated that undergraduates began to worry that if they didn't sign up immediately he might leave and they would miss out. One of those students, Lee Zamir, recalled: "We were told that he could retire at any time, so there was this debate: can you wait to take it when you're ready, or do you just take it because you might not get another shot at it? So I took the class a year earlier than I would have liked to, because I didn't want to roll the dice." A fellow faculty member described Bose as "an artist at making complicated things simple."

A defining moment in Bose's life occurred in 1956, when he bought a pair of expensive loudspeakers for the stereo system in his apartment in Cambridge, and was devastated when the sound they produced was nothing like the sound of a live musical performance—especially the strings. (He'd studied violin when he was young, and had a good ear.) That experience led to a lifelong interest in acoustics and, in 1964, to the founding of Bose Corporation, in an industrial park in Natick. He viewed the company not just as a manufacturer of electronics equipment but also as a basic-research facility similar to Bell Labs, and he viewed MIT as a sort of semiofficial adjunct. The relationship between the company and the university was formalized in 2011, two years before Bose died, when he gave MIT a majority of the company's nonvoting stock on the condition that the university neither sell the shares nor be allowed to participate in the running of the company.

The head of acoustic research at Bose today is Bill Rabinowitz. He grew up in New Jersey, got an undergraduate degree in electrical engineering at Rutgers, and went to graduate school at MIT. The war in Vietnam was going on, he told me, and he knew he didn't want to end up designing missiles, so he took an electrical engineering course called Sensory Communications, which was mostly about hearing. He did hearing-related research at both the University of Illinois and MIT, and did practical work on hearing aids and cochlear implants. Then he got a call from Bose and went to work there, mainly on "projects and problems that have a complicated perceptual element— too complicated for engineers alone."

Dianne Van Tasell had known Rabinowitz for years: they were about the same age and had been graduate students at the same time. "I told him what we'd done with EarMachine," she said. "And Bill told me that the number-one question Bose was asked by its customers was why they don't make hearing aids." Kevin Franck had a Bose contact, too; Lee Zamir, Amar Bose's former student, had ended up working at the company, like many other MIT graduates. In the past, Bose's response to inquiries about hearing aids had always been negative, partly because of the forbidding regulatory environment and partly because Bose headphones had always been of the earmuff type—too large for any modern hearing aid. But federal law seemed increasingly likely to change, and Bose had begun manufacturing smaller, in-ear devices, including noise-canceling earbuds. "They were still a long way from hearing aids, but at least we could imagine someone wearing them for an application like that," Rabinowitz told me. Bose also had a long history of manufacturing sophisticated audio equipment with simple controls, based on observations by Amar Bose and others that even audiophiles seldom made consistently good use of complex controls when they were available. Bose bought EarMachine in 2014, and Franck, Sabin, and Van Tasell became Bose employees. "We thought we could change the landscape of how we got products into people's hands," Rabinowitz said.

I HAD LUNCH WITH FRANCK IN MARCH 2017. Snow had fallen overnight, and the restaurant I'd picked, near my home, wasn't as crowded and loud as it usually is. "That's too bad," Franck said. He was on his way to New York and had made a detour to let me try the Bose product that EarMachine had become, called Hearphones. Hearphones are self-adjustable, high-fidelity headphones designed, in part, to enable people to hear better in noisy places, and they were then in limited release. Franck had asked me to pick the loudest restaurant I knew of, so that we could give them a good test.

In comparison with modern hearing aids, Hearphones look like a technological throwback: a pair of acorn-size earpieces connected by wires to a choker-like yoke, which you wear around your neck. (They look exactly like Bose's QuietControl noise-canceling headphones, which were developed at roughly the same time.) Hearphones seem clunky to someone who's used to hearing aids, but the fact that they're not meant to be concealed gives them a number of advantages. They have room for a large antenna, a big rechargeable battery, high-quality speakers, and more technology than can be hidden inside an ear canal—all features that enable Hearphones to do things that hearing aids can't, while also doing essentially everything that hearing aids can.

I put on a pair. "One of the things you get really good at when demonstrating this device is talking without saying much," Franck said. As he chatted away, I adjusted various settings with a smartphone app that I'd downloaded a few minutes before. The main control was a slider labeled "world volume"—the direct descendant of the loudness dial in EarMachine. "By moving the slider up you can make the world very loud," Franck said. "But it uses that same kind of gain you get in your hearing aids. It amplifies only soft sounds, and doesn't make loud sounds louder, and the reason is that if you're having trouble hearing soft sounds, making loud sounds louder, too, won't help you." I could raise the volume on a scale that ran from 0 to 100, making Franck easy to understand even when he spoke quietly.

One of the best features of Hearphones is that I could slide the slider in the other direction, too—all the way to −50—because Bose had incorporated its justly famous active noise-canceling technology. Moving the slider below zero enabled me to make the entire room quieter—exactly as I would have been able to do with a noise-canceling headset. Such headsets work by analyzing incoming sound waves, then generating sound waves that are identical to them but exactly out of phase—a little like wearing a vest that punches back against any fist that punches you, neutralizing the blow. In 1986, Bose made noise-canceling headphones for two pilots who spent nine days flying an airplane around the world without refueling—the ancestor of noise-canceling headphones today.

People with tinnitus always ask why no one has used the same principle to make tinnitus-canceling headphones. The answer is that it can't be done, and the reason is that the same principle doesn't apply. Sound is the transmission of mechanical energy through a medium; tinnitus is phantom activity within the wiring of the head. Bill Rabinowitz at Bose told me, "With tinnitus, there are no mechanical waves that we can cancel. And, even if there were, we need to have a proxy for what those waves are in order to try to cancel them. With noise-canceling earphones, the things we want to cancel arrive at the device, so we can sense what they are and try to figure out how to cancel them. If we have no access to what it is we're trying to cancel, we can't sit there and try to create waveforms and adjust them—that's a hopeless concept." Sad to say, noise-canceling headphones often actually make tinnitus seem worse, by turning down the volume of background sounds that would otherwise mask it.

The noise-canceling principle can be used to cancel other kinds of vibrations, however. In the 1980s, Bose experimentally built automobile suspensions that enabled a driver to drive over speed bumps without feeling the bumps. (Search YouTube for "Bose active suspension.") The suspension technology—which used a microprocessor, magnets, and servo motors to make a moving car feel as though it were always traveling straight over level ground—was too expensive to implement commercially, and in the decades since then it's been made mostly redundant by the development of such now common automotive features as computer-controlled suspensions. But the company adapted the same bump-canceling idea to truck seats, which it still sells, through a division called Bose Ride. The seats are expensive, but fleet operators have found that reducing the vibrations that truck drivers have always endured, day after day for hours at a time, increases productivity and reduces healthcare costs.

Franck went on. "Once you've found a volume that feels right to you, move over to the slider on the right. That fine-tunes it. As you slide it down, you get more bass and the sound is fuller; as you slide it up, toward treble, you hear more consonants." A button at the bottom of the screen allowed me to select among three directional settings. I could focus narrowly on Franck, who was sitting directly across from me, or I could widen the range to 180 degrees, or I could bring in the entire room. "Right now, you're in the 'focused' mode," he said. "So the sounds that are coming from where you're looking are louder than the sounds coming from everywhere else. In fact, there's so much directionality that if there were people sitting to your right and left, you wouldn't be able to hear them as well as you can hear me. But you can bring those people in by switching to the mode called 'front.'"

I did that. Then I switched to "everywhere," and without turning around I could now hear things happening behind me—mainly some chefs and waiters talking to one another and banging stuff around in the kitchen. I was able to play music in the background as we conversed—with far better fidelity than is possible with even the most expensive wireless hearing aids, because Hearphones have better speakers and their big battery makes it possible for them to use a more robust and power-hungry version of Bluetooth—and I could raise and lower the music's volume independently from everything else. If my phone had rung, two directional microphones inside the earpieces would have automatically aimed themselves toward my mouth when I answered it. The app also comes with four preset modes: focused conversation, group conversation, music, and television. I could have used any of those instead, or tweaked them by adjusting their saved settings, and I could have added as many as ten preset modes of my own creation. I was also able to adjust the right/left balance, and even to mute one ear entirely.

"Here's the story we tell people," Franck continued. "Imagine you're at your house, and you're going to walk to the train station to get to work. On the way, you want to listen to a podcast, so you turn it on. You also turn the world volume on your Hearphones up and you switch the focus to 'everywhere,' because in addition to the podcast you want to be able to hear cars and everything else going on around you as you walk to the station. When you get on your train, though, you turn the world volume all the way down, to minus-fifty—and now the podcast is the only thing you can hear. You get to work and go to a meeting, so you turn the world volume back up and adjust the focus. And you don't need the app to do any of that; there are also controls on the device itself. If you want to, you can just kind of park your Hearphones around your neck all day, and when you're not using them you can take the earbuds out of your ears and let them hang or tuck them under your collar." A significant advantage of wired earbuds over wireless earbuds is that you can remove one or both without putting them in their little case and then forgetting what you did with the case.

Hearphones, legally speaking, are not hearing aids. But they use the same microprocessor that hearing aids do, and, although they aren't promoted that way, they address the same issues. Because they have been optimized for performance rather than for concealment, they have a number of significant advantages, not least among them cost. (I paid five hundred dollars for mine.) Their many fascinating features include noise cancellation inside the ear canal. When you fully block an ear, with a silicone earplug or an earbud or an earmold hearing aid, you turn the ear canal into an echo chamber. "If you plug your ears and talk, or even just walk, you hear boom, boom, boom," Franck said. The booming sounds are low-frequency noises that you yourself have made—by talking, walking, chewing—and that have reached your ear canal by bone conduction, through your skull. "What happens is that those sounds get into your ear canal but can't get out," Franck continued. "That's because the vibrations are hitting your eardrum and just kind of resonating, and it can drive you crazy." (This is the so-called occlusion effect, or head-in-a-barrel effect, which I described in chapter seven.)

Hearing-aid manufacturers address occlusion by creating a tube-like vent in the hearing aid—an escape route for the trapped sound. This can work well for users who don't need much amplification in the lower frequencies, as is the case with most people who have hearing loss. But the vent allows high-frequency sounds that are amplified by the hearing aid to escape, too, and when those sounds reach the microphone on the external part of the hearing aid they can create feedback. "Audiologists are constantly balancing the problem of occlusion with the ability to provide gain where you need it," Franck said. "They'll ask you how much the occlusion bothers you, and if it bothers you a lot they'll make a bigger hole in the earmold—but once they've done that they can no longer give you the gain you need in the higher frequencies. It's like Whac-A-Mole with patient complaints." He said, furthermore, that Bose's historical success at dealing with occlusion was among the main reasons that he, Sabin, and Van Tasell had been eager to make a deal for EarMachine.

When Hearphones were in development, Bose held focus groups at its headquarters. "We made some announcements on Facebook, and we invited people who lived within seventy-five miles of Framingham and had bought Bose products before," Franck said. "We told them that we wanted them to try something that would help them hear better in noisy places." They turned a basement meeting room into a fake café, piped in sounds recorded in a bar, and turned up the noise. They also created a fake TV room, and showed the testers that they could each use their Hearphones to bring the volume to a personally comfortable level without touching the television itself. (Disagreements about how loud the TV should be are a known source of marital stress.)

"We spent an hour and a half with these people, listening to what they liked and didn't like about the product," Franck said. "We wanted to be sure that people could understand how to use them, and that they wouldn't be calling customer support and treating them like an audiologist, because that wouldn't scale. We also wanted to make sure that we weren't confusing anyone, from a regulatory point of view, into thinking that Hearphones were a hearing aid."

Dan Gauger, a longtime Bose engineer, was involved in the decision to acquire EarMachine and develop Hearphones. We met two months after my lunch with Franck. He told me, "I have a friend who's in his late sixties and has been wearing hearing aids for six or seven years. Unlike your average consumer, he's passionate about sound and hearing, and fanatical about the way things should sound, and he talked an audiologist into giving him a copy of his fitting software, so that he can tune his hearing aids himself." Gauger and his friend attended a conference while Hearphones were still in development. "I had a prototype, and he used them over dinner one night, with a bunch of other people in a loud restaurant, and then the next day he used them over lunch in a hotel restaurant that was much quieter. Then he dragged me into an unused conference room, which was dead quiet. In each of those places, he compared Hearphones with his three-thousand-dollar-per-ear hearing aids that he had personally tuned. And afterward he said to me, 'In the quiet conference room I slightly prefer my hearing aids, but everywhere else you guys beat them hands-down.'" A college friend of my wife's and mine, who lost a significant amount of hearing to Ménière's and learned about Hearphones from me, told me in an email, "They are great for me in restaurants and other noisy environments. A funny/sad observation: my audiologist doesn't want to hear that they are better than my hearing aids in noisy environments, because 'they shouldn't be.' She's part of the system!"

I VISITED BOSE ITSELF IN 2018. The company's headquarters are right next to the Massachusetts Turnpike, twenty miles west of Boston, on top of a big hill called the Mountain. The complex is virtually across the street from the Sheraton Framingham, a hotel that, as you whiz past it on the Mass Pike, looks a little like a medieval castle and a little like a half-timbered Tudor mansion. (From the inside it looks disappointingly like a normal hotel.) I met Bill Rabinowitz—Dianne

Van Tasell's old friend and one of the people involved in the company's decision to purchase EarMachine—in his office. He was wearing jeans and a checked shirt, and on the screen of his computer was a hearing-related scholarly paper whose title I couldn't make out from where I was sitting.

I asked him about his own hearing. "It's about appropriate for my age," he said. "I've got some noticeable high-frequency loss. In a situation like this, talking to you in a quiet room, I don't really need any help. But if I go into noisy environments or lecture halls I notice that, my god, some of these people speak so softly, and there's reverberation. I wear Hearphones in some of our meetings, but I haven't really worn them outside the company, yet, except at restaurants with my family. I've taken a bunch of hearing tests. My wife is an audiologist. I met her on a research project. She's always telling me that my hearing is awful. I think hers is worse, but she doesn't believe that."

He took me to see the Bunker, a room in which Bose tests the durability of new speakers. "Loudspeakers are crazy devices," he said. "Ultimately, they have to move—they have to move air—and things that move are a pain in the ass, because if you don't design them carefully they can destroy themselves. So we try to kill them before they get to customers, by playing the worst-case signals 24/7, for weeks. It's like a torture chamber." We entered a room that contained racks of electronic equipment. The air was throbbing.

"This isn't it," Rabinowitz said. "These are just the amplifiers." There was a large selection of over-the-ear hearing protectors, and he told me to find a pair that fit well. "If you're wearing your hearing aids, take them off," he said. "I don't think this would damage your phone, but you might want to leave it here, on the table, just in case."

In a thick concrete wall was an orange-painted steel door covered with dire warnings (no children; no people with heart problems). He opened it, and we stepped into an airlock, and he told me to pull the orange door tight behind us. Then he opened another steel door, and we entered the Bunker itself.

Even though I was wearing industrial-strength earmuffs, the room felt life-threatening—undeniable proof that sound is a physical force. Tom Krehbiel, an audio critic, once visited the same room and reported that he "had the feeling that I was undergoing unrelenting CPR from a thousand tiny hands." Rabinowitz had me place my hand on a big speaker, which was playing a very low-frequency sound at an extremely high volume; the surface was vibrating so powerfully that touching it was painful. Another speaker, playing a very high-frequency sound, also at an extremely high volume, was hot to the touch. We went back through the airlock after a minute or two; I had no desire to linger.

Rabinowitz then took me to another test room, built by Amar Bose in the 1980s. It's called the Reverb Room, and it's the opposite of an anechoic chamber. The floor, walls, and ceiling are thick, as in the Bunker, and they're highly reflective of sound, and none of them forms a right angle with any of the others. Rabinowitz told me to imagine that the walls were mirrors. If he shined a laser at one of the mirrors, he said, the beam would bounce off obliquely, like a caroming billiard shot, and it would bounce around so much, from one angled surface to another, that the entire room would fill with light. The angled surfaces, he said, were doing the same thing to sound waves.

"This room creates an isotropic, or diffuse, sound field, which approximates what it's like to be in a space in which sound is coming at you from all directions equally," he said. "It's like an extreme version of a restaurant, or a highly reverberant church." Among other things, Bose uses the Reverb Room to test noise-canceling headphones. On a table were many pairs—along with a small round makeup mirror on a stand, for checking fit and appearance. "Noise cancellation is a lot harder than most people think," he said. "It's not so difficult to make a noise-canceling device if all the sound is coming from just one direction. But if you're a passenger in an airplane, or you're sitting in Starbucks, the sound is all around you. So when we design noise-canceling products we want them to work in this environment." The Reverb Room was almost as unpleasant to hang around in as the Bunker: my head felt like it was under assault. Another Bose scientist told me that members of the Grateful Dead's road crew had once been given a tour of the Reverb Room, and one of them had said, "It sounds just like Boston Garden."

HEARPHONES ARE CLASSIFIED BY THE FDA as personal sound amplification products (PSAPs), which it distinguishes from hearing aids. According to guidance issued by the agency in 2009, a hearing aid is "a wearable sound-amplifying device that is intended to compensate for impaired hearing," while a PSAP is "a wearable electronic product that it not intended to compensate for impaired hearing, but rather is intended for non-hearing impaired consumers to amplify sounds in the environment for a number of reasons, such as for recreational activities." Four years later, the FDA reaffirmed that distinction, and warned PSAP manufacturers not to suggest in any way that their products were intended to solve problems commonly associated with hearing loss, including "difficulty listening to another person nearby, difficulty understanding conversations in crowded rooms, difficulty understanding movie dialogue in a theater, difficulty listening to lectures in an otherwise quiet room, difficulty hearing the phone or doorbell ring, or difficulty listening in situations in which environmental noise might interfere with speech intelligibility." So Bose, in its promotional materials for Hearphones, doesn't talk about hearing loss at all; it emphasizes their noise-canceling capability and describes them as "conversation-enhancing headphones that are specially designed to help you hear in louder environments" and that make "any conversation in a noisy place easier and more comfortable."

PSAPs, or products like them, have been around for a long time, and some of them are junky. But in recent years several manufacturers have introduced models that are both effective and affordable—so much so that the National Academy of Sciences and the President's Council of Advisors of Science and Technology have said that PSAPs can be sensible choices for people who have relatively mild hearing loss. The best ones contain the same technology that hearing aids do, yet cost much less. There's always a risk that you'll buy something that you end up not liking. But that happens with hearing aids, too, and with hearing aids mistakes are more costly.

Not long after the introduction of Hearphones, Kevin Franck left Bose to become the head of audiology at Mass. Eye and Ear. (Franck was the audiologist, mentioned in chapter seven, who played those garbled audio files of Homer Simpson and Barney Fife for me.) I visited him in his office there in the spring of 2018, and he showed me a selection of PSAPs he liked that were available at that time—beginning with EarMachine, the Hearphones predecessor, which you can still download, for free, as long as you have an iPhone. "The only problem is that you have to have headphones that are connected by a wire, so you also need a headphone jack," he said. The reason is that if you used wireless ones you'd notice a disturbing lag between movements of a speaker's mouth and the voice you heard—the result of a Bluetooth limitation known as latency, which is the delay caused by the time it takes to send, receive, and process the radio signal. (Latency isn't an issue with Hearphones because their microphones are in the earbuds, not in your Bluetooth-connected phone.) "But if we're at dinner I can push my phone across the table, and now I hear great, because the microphone is close to you, and I've got all the fidelity that I'm used to with my phone."

Next, he showed a device that worked in a similar way but didn't require a smartphone: one of several SuperEar sound amplifiers from Sonic Technology Products, a California company that, in addition to PSAPs, makes detachable trays for airplane window seats and devices that emit ultrasonic signals intended to repel rodents and other pests. There are three SuperEar models, and, although they aren't free, they're cheap: $50, $60, and $80. "All you have to do is be willing to wear headphones," Franck said. Each model comes with two different kinds of headphones, one less conspicuous than the other, along with various accessories. The receiver is small enough to carry in a shirt pocket. You place it on the table in front of you or attach it to your belt, and the controls on the receiver allow you to make adjustments. At Franck's direction, Massachusetts General Hospital now asks all inpatients, at intake, to rate their own hearing—a sort of self-triage. Instructional posters are hung above the beds of those who have hearing issues, and the ones who have hearing issues but don't have hearing aids are given SuperEar devices. The posters remind doctors and nurses to speak up and to turn off TVs and other sound sources, and to ask patients who need hearing devices to turn them on. Patients can't follow their doctors' instructions if they can't hear what those instructions are or if they misunderstand what they think they hear, and doctors routinely underestimate the severity of their patients' hearing problems. "Fifty dollars is the same cost as a bag of saline," Franck said. "If you're willing to wear paper clothes, you're going to be willing to do this—right? And our hope is that, once you've realized that it lets you hear your doctor better, maybe you'll also realize that your ears truly are a problem, and once you're out of the hospital and feeling better you'll look for a hearing solution." (Other hospitals, among them those at Johns Hopkins and the University of Pittsburgh, have instituted similar programs.) Medical personnel with hard-of-hearing patients can wear surgical masks with clear plastic windows over the mouth—a boon to lip-readers— although COVID-19 has made such masks hard to find.

He then showed me a CS50+, a PSAP from Sound World Solutions, which is based in Illinois. It looked a little like a conventional behind-the-ear hearing aid and a little like the kind of Bluetooth telephone earpiece that limousine drivers often wear. "This is the battery," Franck said. "It's rechargeable, and it snaps on, and it goes behind your ear. Microphone here, microphone here—just like hearing aids that have two microphones. You've got to download their app and make your own adjustments. Three hundred and fifty dollars." The CS50+ blocks the ear canal, so occlusion can be an issue, especially if you wear two. "A company called Nuheara sells PSAPs that look like earbuds," Franck continued. "They're about three hundred dollars, but with them I find the occlusion to be pretty hard to deal with. Bose Hearphones are unique in having noise canceling to address occlusion. To me, that's the jewel of that product." (I have the same Nuheara earbuds that Franck showed me, and I agree about the occlusion.)

There are other PSAPs as well, and by the time you read this there will be more. "What I love about all of these devices is that, because of them, people with a bit of hearing loss can begin to enjoy the benefits of the type of signal processing that used to be available only in hearing aids, without spending thousands of dollars," Franck said. Once users have been convinced of the basic benefit, he continued, an audiologist can help them make further gains, if they still have issues. "Maybe you're getting feedback and you need a custom earmold to get rid of it—but then you get occlusion, so now you need a vent. Or maybe you decide you want something that's invisible. The further you go, the more trade-offs there are, and to balance them you need a person who is trained in all this stuff." As a low-risk introduction to the field, he said, PSAPs are ideal. And they may be all that people whose hearing loss is neither severe nor unusual will ever need.

BOSE, BY LAW, can't call Hearphones hearing aids. But I can. And, as far as I'm concerned, they work better than my hearing aids for everything that my hearing aids are supposed to do—and with greater fidelity and no annoying hiss. I use them as my regular headphones, and I take advantage of their noise cancellation if my wife, who sings in a church choir, is practicing in the kitchen while I'm trying to work. I once wore them in a sports bar with a group of my-age golf buddies (who passed them around the table), and I wore them recently when my wife and I went out to dinner with two other couples. And I wear them to movies, where, if I turn up the volume before the lights go down, I'm amazed at how clearly I can hear people in distant parts of the theater chewing popcorn and making boring conversation. Actually, movies nowadays are often so loud I'm more likely to turn the volume down than to turn it up. The only negative is that if I pull out my phone to make adjustments once the movie has begun, a stranger sitting near me will think I'm checking my news feed and hiss at me to put my damned phone away. (I've learned to cover it with a jacket while I fiddle. The headset has built-in controls, but they're small.)

In 2017, Congress passed the Over-the-Counter Hearing Aid Act, which required the FDA to create a new category of hearing aids for people with mild to moderate hearing loss, enabling them to buy hearing aids without help from an audiologist. The Senate version was cosponsored by Elizabeth Warren and Chuck Grassley, and the vote was 94–1. (The only senator to vote against it was Bernie Sanders, whose objections had to do with a different issue involving the FDA.) A critical factor in passing the legislation was an intensive educational effort by the Hearing Loss Association of America, which is based in Bethesda, Maryland, and has chapters all over the country. (Among many other things, the HLAA and its chapters hold self-help meetings for people with hearing problems, and provide numerous hearing-related services.) Many individuals were involved as well. "People were just pissed at how much hearing aids cost," Dianne Van Tasell told me. "I would go into Congressman So-and-so's office and say, 'Do you have a relative who has a hearing loss, who has tried to buy hearing aids?' And usually it would be the mother, and the cost would have been something like six thousand dollars. And then I would ask, 'Had you known that Medicare wouldn't pay for them?' And they'd say 'No!' and I'd say, 'Did you know that it costs the manufacturer less than a hundred bucks to make them?' That's why this bill had such broad bipartisan support. The hearing-aid industry is screwing consumers, and it just isn't right.'" The next big battles will be over gaining coverage for hearing aids under Medicare and preventing audiologists and the major hearing-aid manufacturers from persuading the government to place low limits on the power of the over-the-counter aids. "The hearing-aid industry and the audiologists are looking to hobble the OTC devices, so that people will put them on and say they're not helpful," Franck told me. "Their attitude is, 'Fine—let them have toys.'"

In late 2018, the FDA approved Bose's request to sell "the first hearing aid authorized for marketing by the FDA that enables users to fit, program and control the hearing aid on their own, without assistance from a health care provider." The Bose hearing aid was approved for "individuals 18 years or older with perceived mild to moderate hearing impairment (hearing loss)," who, the FDA said in a press release, would be able "to fit the hearing aid settings themselves, in real-time and in real-world environments without the assistance of a health care professional." In some states, purchasers will still need to make their purchase through a "licensed hearing aid dispenser." But that will probably change, too, once the FDA has finalized its rules for over-the-counter hearing aids. The market will be chaotic for a while, but in the end there will be more options for more people. In an ideal world, the pricing would be transparent, and audiologists and other providers would offer a range of solutions, including but not limited to conventional hearing aids: the kinds of options that Franck demonstrated in his conversation with me. A helpful step toward full transparency would be to "unbundle" the cost of hearing solutions, by breaking down prices into their components: so much for an evaluation, so much for a device, so much for a return visit—or any other reform that would make buying a hearing aid feel less like buying a used car.

The hearing-aid development team at Bose wouldn't give me specifics about the price or the date of availability, but their product will almost certainly be for sale by the time you read this, and similar products from other companies will join it soon, if they haven't already.

You can already buy hearing aids over the counter, sort of. Audicus sells good, relatively inexpensive hearing aids online, and ships them by mail—but if you need adjustments or repairs, you have to mail them back. That arrangement is not ideal, but more options are coming. Sabin told me, "I think it's possible that, in the future, a hearing aid will be a feature, not a device. To make a hearing aid, you need a microphone, a processor, and a speaker, and the number of devices that have those things has gone up exponentially in the past few years—with Bluetooth headphones, AirPods, all kinds of wearables. If you want to make a *good* hearing aid, you have to be sort of deliberate in how you assemble those components, but if you want to make an *O.K.* hearing aid, you can simply add that functionality to a headphone, a voice assistant, a fitness tracker. The number of options, especially affordable options, already has increased, and it's going to keep increasing, especially over the next five or ten years."

Ten

COCHLEAR IMPLANTS

The biggest advance in hearing technology in recent decades has been the development of cochlear implants. These are electronic devices that digitally process speech and other sounds in something like the way that hearing aids do, but then convert the signals into electrical impulses, which are conveyed to an array of electrodes threaded into the spiral of the cochlea. Cochlear implants directly stimulate auditory nerve fibers, thereby bypassing the cause of the recipient's hearing problem, and the brain interprets those stimuli as sounds. Part of the device is anchored to the skull, under the skin, near the ear; part attaches magnetically to the anchored part, through the skin, and usually protrudes through the hair; part is deeply hidden within the temporal bone and the winding innermost channel of the inner ear; and part hooks over the ear and resembles a hearing aid.

Experiments with cochlear implantation began in the late 1950s and early 1960s, in France and California; the first versions of the modern technology were developed during the following decade. One of the pioneers was Graeme Clark, an Australian professor of otolaryngology, who at one point tested an idea he'd had by pushing a blade of grass into the curving inner channel of a seashell he'd picked up on a beach in New South Wales. The performance of cochlear implants has improved steadily and dramatically since then. A scientist who worked on a recent major research project told me, "In many ways, cochlear implants are the largest, most dramatic functional replacement of a sense, by far. There is nothing equivalent in vision. People are working on retinal implants, and things like that, but they are still very primitive." A professor of otolaryngology told me, "I think that, of all the things we've done in otology in the past twenty-five years, cochlear implants are the most spectacular. They're undeniably the finest biological prosthesis that we have today, for anybody, in terms of restoration of function."

THE RADIO PERSONALITY RUSH LIMBAUGH lost his hearing to an autoimmune disorder (and perhaps partly to the drugs he received as treatment for that disorder) in late middle age. He was implanted on the left side in 2001, when he was fifty, and on the right side in 2014, after he'd become convinced that no medical cure for his type of deafness was likely to be found in time to do him any good. In a radio broadcast in 2014, a transcript of which is available on his website, he spoke at length about his experience, and also about watching someone else being implanted the week before his own operation. He described the surgical technique as "sculpture of the skull" and said, "Eighty percent of this is a high speed drill, the surgeon using a high speed drill like a dentist, and just carves out, sculpts a place in the skull for the implant to go. You cannot drill straight down because you don't want to go to the brain. You gotta stay just short of that, so you drill down at an angle and the implant is about, oh, I'm thinking of trying to give the shape, a bell shape that's about two inches long and maybe an inch and a half at its widest and a half inch at its narrowest. They have to sculpt a trench for it, and then they have to sculpt a canal from that to the cochlea in the ear. They connect it and they take tissue from another part of your body to connect it, and then they sew you up."

Hearing people usually assume that cochlear implants enable deaf people to hear the way hearing people hear, a belief that's supported by online videos of recipients reacting with joy as their brand-new devices are turned on for the first time. In one video I watched recently, on YouTube, a young mother bursts into tears when her year-old daughter, sitting in her lap and sucking a pacifier, reacts to sound and reaches for her ear. Watching it made me cry, too. But the video is brief, and when I watched it again, later, I decided that the baby's behavior, in contrast to the mom's, was ambiguous. An audiologist who works with implant recipients of all ages told me that the average infant's reaction to implant activation is more likely to be tears, perplexity, or nothing.

Implants don't function like undamaged ears. Limbaugh again: "It's impossible for me to describe, or anybody that has a cochlear implant, it's impossible to describe what things sound like. It's totally artificial because in my memory of hearing there isn't anything I ever remember hearing that sounds like the way I hear things now. The closest that I could come to it—and this doesn't get there, but, I mean, this is the closest in trying to help people understand how I hear things is scratchy, static AM radio. That's not it, but that's as close as I can get." Music, for Limbaugh, no longer exists. "I don't have the frequency response to identify melodies, even music that I've heard," he said. "My memory supplies the melody. I can turn on one of my favorite songs from the seventies, if I didn't know what the song is, if I don't have a piece of text or if I don't have somebody tell me, I will not recognize it. I need to know what it is. When I do, then my memory supplies the melody and the lyric and I can hear it. But, if I don't know what it is, it's just noise of the same note. Music in a movie, the sound track to a movie sounds like fingernails on a chalkboard."

Quality of outcome from cochlear implantation depends on many factors, among them the age of the recipient at the time of implantation and how long the recipient has been deaf. Among adults, recipients who had fully functioning ears for most of their lives do much better than those who have never heard anything, because they remember what sound sounds like, as Limbaugh does, and they already know how to speak. In addition, their auditory circuitry is usually in better shape.

Limbaugh, therefore, was in many ways an ideal recipient. He had lost his hearing both suddenly and recently, and when he went deaf he'd had half a century's exposure to the world of sound. Still, he has had to make significant adaptations. His implants have enabled him to hear well enough to continue to be a public nuisance on the radio and elsewhere, but he needed speech therapy to make his voice sound semi-normal, and he's had to learn to read lips, and when he's on the air he relies partly on real-time transcription of whatever the people who call in to his show are saying, to fill in what he can't hear on his own. Even so, he views cochlear implants as an astounding innovation, and knows that without them his radio career would have been over two decades ago. And there is no question that his quality of life is vastly higher than it would have been if he'd lost his hearing two decades before that, when the technology was rudimentary, or if he hadn't been implanted at all.

PEGGY ELLERTSEN IS SEVENTY YEARS OLD. She majored in speech pathology at Northeastern, and during her senior year the head of her department received a grant to create an audiology clinic on campus. She volunteered to be tested, so that he could experiment with the new equipment, and to her surprise the tests revealed that she had a mild hearing loss. She was fitted for hearing aids shortly afterward, but she hated the way they felt and sounded. She put them away, as many first-time users do, and didn't try them again until she was in her thirties and doing graduate work in speech pathology. "I didn't know anyone my age who had hearing loss," she said. Nevertheless, she gradually became accustomed to thinking of herself as a person with a hearing deficit. Her professional specialty was child literacy, but as her hearing deteriorated she shifted her focus to the oral rehabilitation of adults with acquired hearing loss. She attended a convention of the Hearing Loss Association of America—a life-changing event for her, because it was the first time in her life that she'd been exposed to a community of people who also had serious hearing problems. She became active in the HLAA, and in 2015 she was elected to its board of trustees. She has been certified as a hearing-loss mentor through a program at Gallaudet University, and she has worked intensively with a therapist who showed her how to boost the effectiveness of her hearing aids by learning to "laser focus" on what people around her were saying.

I met Ellertsen in an examination room at Mass. Eye and Ear in 2018. She'd had a cochlear implant on her right side a little over a month before. The implant had been activated about three weeks after that, and she was returning to her audiologist for adjustments and a one-week checkup. (There's always a delay between implantation and activation, to allow incisions to heal and swelling to go down.) She said that she had just had a haircut, the better to show off her device. The visible parts were a pearly goldish color—the same as the hearing aid she wore in her other ear, and similar to the color of her hair.

Her audiologist, Sarah Laurello, asked her what she thought of her implant so far. "It's not perfect, but I don't care," she said. "It's turned out to be easier than I thought it would be. What I'm finding is that, especially when I'm in a car and listening to interviews on the radio, I now know what people are saying—I actually know what they're talking about." She said that she and her husband had had a Skype conversation the night before with their daughter and son-in-law, who live in New Orleans, and that when it was over she hadn't had to ask her husband to recap what everyone had said—for her, a first.

Ellertsen had waited several years longer than she might have before being implanted, she said, mainly because she'd been worried about potential side effects—especially vertigo. Inner-ear surgery always carries risks, including damage to the vestibular system, diminished sense of taste, and facial paralysis. "I kept saying that I didn't want to trade one problem for another problem," she continued.

"My expectation was that when it was over I was going to be staggering across the room, and that my husband would never be able to leave the house because I would be done in by vertigo. But that hasn't happened."

I asked her what her implanted ear sounded like.

"It's a little bit electronic, a little bit robotic," she said. "People say it sounds like Mickey Mouse, or a chipmunk. But that's already starting to fade. I spent about three hours with two women yesterday. We had lunch, and we were talking, talking, talking, and I realized at the end of that time that they no longer sounded like chipmunks." She said that one thing that did bother her was low-level ambient noise, like the sound of traffic or of a ventilation system, which she perceived as "a constant high-pitched wind." She also said that she was having more trouble than usual deciphering speech in noise.

"That is something that will definitely develop over time," Laurello said. "What we've essentially done on your right side is give you access to a lot more sounds, and it's going to take your brain a little more time to figure out what you want to pay attention to." Suddenly hearing after years of near silence is jarring, no matter what the quality of the sound, and recent recipients often feel overwhelmed. (This happens to people with new hearing aids, too. A hearing aid that sounds pleasant the first time it's worn isn't necessarily doing everything it ought to.) Overall, though, Ellertsen was pleased. "People say, be careful, don't expect to be a rock star, this is going to be really hard," she said. "But it's not that hard. This is so much less hard than other things I've done in my life. This is like a slightly complicated dental procedure, with lots of Novocain."

A FEW MONTHS BEFORE Ellertsen's appointment with Laurello, I had spoken with Meaghan Reed, who is Mass. Eye and Ear's associate director of audiology. Reed told me that she had become interested in her field in college, in Florida, following a visit to a school for the deaf and blind, and that at one point she had worked in an ear-nose-and-throat medical practice, whose older patients mostly had age-related hearing loss, and whose younger patients mostly had ear infections. I sat in a chair near her desk, and, as we were talking, I could hear what I suddenly suspected might be a secret hearing test. "Is there a ticking clock in here somewhere?" I asked. She said there wasn't—but then she realized that there was.

"The clock doesn't work, although I guess it really does tick," she said. Presumably, she had become so accustomed to it that she no longer registered its existence. This is a well-known trick of the brain, which has evolved to ignore steady, monotonous inputs, the better to notice unexpected threats—a sudden movement in the underbrush, the ominous crackling of dry leaves. As a result, we become habituated to the steady hum of distant traffic, the whir of our computers, the unceasing din of modern life, the hiss of our tinnitus. The office of a former literary agent of mine had a tall dropped ceiling, within which, somewhere, was a smoke alarm whose battery had run down. The regular beeping drove me crazy during my (infrequent) visits, but the agent and her assistant were so used to it that they no longer heard it, unless someone (me) pointed it out.

Part of Reed's job with cochlear-implant patients is managing expectations. "We counsel them that, on Day One of activation, their brain is not going to understand what they hear," she said. "For some people it sounds like a tone; for some, it sounds like a *boop*, a *beep*; for some, it sounds like Charlie Brown's teachers. Some people say, 'Yeah, it doesn't sound normal, but I can hear speech and I can understand that somebody is saying something.'" Others aren't certain that the sounds they now hear are even potentially intelligible. "Sometimes we can predict what the result will be, but sometimes two people who have what seems like the same history and experience will have totally different responses."

New recipients are tested, in part, by being placed in a sound-isolated booth and asked to identify monosyllabic words. I asked Reed why they didn't use complete sentences. "With monosyllables, you get a better sense of what's missing," she said. "You don't give them a song; you give them a note. And if you do give them a sentence test, you give it against a background of noise." The purpose is to eliminate clues from context. Reed said that performance on the monosyllable test continues to improve for about a year or a year and a half after activation, and that most of the improvement comes during the first three months, and that results cover a broad range. "After a week, some people are doing extremely well, while others are barely getting pattern perception," she said. "The performance is pretty variable."

AFTER ELLERTSEN, LAURELLO, and another audiologist had talked for a while, we moved down the hall to an audiometric testing room. Ellertsen sat in the booth, and Laurello gave her the same kind of hearing test I'd had at Starkey, both with the implant turned on and with it turned off. The tone test showed, among other things, that

Ellertsen had retained most of the (very limited) hearing she'd previously had in her implanted ear—an unusual result, because the surgery itself often wipes out anything that's left. I asked Laurello why Ellertsen had kept hers. "Just different surgical techniques," she said. "And sometimes the pathology of the ear makes a difference. For example, if there's any ossification, the insertion can be more traumatic." Ellertsen said later that her surgeon had told her that he had bathed her cochlea with steroids before inserting the electrode, in the hope of protecting what she had, but that the preserved residual hearing was so minimal that she wasn't certain it was adding anything to what the implant provided. And Laurello said that Ellertsen wouldn't necessarily keep what she now had, because the electrical impulses from the implants themselves are suspected of doing damage.

Laurello then gave Ellertsen three word-recognition tests—first with no devices; then with just the implant (while a masking sound was piped into her other ear, to prevent the functioning ear from confusing the results); and, finally, with both the implant and the left-ear hearing aid. She made many fewer errors when she was wearing her devices but missed words in all three tests. In one session or another, she heard *fall* as *borrow*; *tough* as *park*; *patch* as *bar*; *white* as *broke*; *hen* as *low*; *course* as *goof*; *yearn* as *warm*; and *got* as *duck*. Something I hadn't realized about such tests is that the scoring is based on recognized phonemes, not entire words. Thus Ellertsen received partial credit for hearing *tooth* as *deuce*, and in theory could have scored more than 50 percent without correctly identifying even one entire word. As it was, her score in the final test, with both devices, was 52 percent. That was 8 percentage points higher than before her operation, but it was much lower than I would have guessed beforehand, based on how easily we had conversed in the examination room. If we hadn't been sitting in an audiologist's office and discussing her cochlear implant and her hearing aid, I wouldn't have suspected that she had a hearing problem—powerful evidence of the contribution of context to comprehension, and also of how hard Ellertsen has worked over the years to improve her listening skills.

We returned to the examination room, and Laurello made some volume adjustments in the processor on Ellertsen's implant, by using her computer and a device that clicked onto the unit. The implant was from Advanced Bionics, which is based in California, and in 2009 was bought by Sonova, the Swiss company that manufactures Phonak hearing aids. The Advanced Bionics device is one of three that the FDA has approved for use in the United States; the other two are made by Cochlear Limited, based in Australia, and MED-EL, based in Austria. Ellertsen had arrived for her appointment carrying the big, colorful box that her earpiece and various accessories had come in, and she and Laurello sorted through various items inside it. An accessory she was thinking of adding is the T-Mic—an Advanced Bionics attachment that moves the implant's microphone from the external earpiece into the opening of the ear canal. Its purpose is to improve speech comprehension in noisy environments by taking advantage of the mild ear-trumpet effect of the pinna. It also enables people with cochlear implants to eliminate an annoyance that also affects many hearing-aid users: the fact that in order to hear a telephone conversation they have to hold the receiver above their ear, so that the phone's speaker is positioned directly over the device's microphone. A T-Mic (the *T* stands for "telephone") enables a user to hold a phone conventionally—and also to use over-the-ear headphones. The same trick isn't available to hearing-aid users, because placing an aid's microphone and speaker inside the ear, right next to each other, would create an unsolvable feedback problem.

THE FDA INITIALLY APPROVED cochlear implants only for adults, but research has shown that, for children who are born deaf or who become deaf in early childhood, the devices are vastly more effective if they're put in before the parts of the brain that process speech have developed fully. Many scientists believe that there is a period, beginning at about twenty months and lasting for perhaps eight or ten years, when our brains are able to acquire language easily—and one piece of evidence for this hypothesis is that people who learn a second language in early childhood are more likely to speak it without an accent than people who learn it in adolescence or later.

What is clear is that people who are born deaf, or who become deaf before they've begun to speak, have a harder time learning spoken language than people whose hearing is fully functional from birth, and they receive less benefit from cochlear implants (or hearing aids) than very young recipients do. The surgery is now sometimes done on children who are six months old—although Meaghan Reed told me that it's usually better to wait until they're ten months or a year, "because we want everything to grow and develop a little bit more first." She also said that infants who can hear virtually nothing will often be fitted for hearing aids before they're old enough for the surgery, because even minimal levels of acoustic stimulation appear to improve outcomes later.

James Henry—the former rock guitarist whose daughter's congenital deafness inspired him to earn advanced degrees in audiology and behavioral neurology, and to become a tinnitus researcher for the Veterans Administration—told me, "Cochlear implants are amazing, but you need to get them when you're young to really benefit from them. If you get them when you're older, your brain doesn't have the same plasticity, and it can't adapt as well to the signals that come from the implant." His daughter, who is now about forty, was born with no hearing. "She has a cochlear implant, but she didn't get it until she was twenty years old, because they weren't fitting children back in those days," he said. "Now they are, and, in fact, my deaf daughter has a deaf daughter, who was fitted for cochlear implants in infancy." His granddaughter had minimal hearing in one ear when she was born, but lost it so quickly that she, like her mother, can be considered to have been born deaf. "The difference between my daughter and my granddaughter is that my daughter had great difficulty learning speech skills," Henry said. "But my granddaughter can hear things and repeat them back without looking at the person who is speaking." This is a remarkable achievement, because Henry's granddaughter, unlike Rush Limbaugh, has never really heard speech, or anything else, except through her cochlear implants. Another scientist told me that trying to learn spoken language under those conditions, even for someone who begins in infancy, is as difficult as trying to learn a language against a background of noise.

Still, the brain often finds a way. Meaghan Reed told me, "If you start stimulation and rehabilitation early enough, with implants or hearing aids, children can catch up before school age—and a lot of them catch up by two to three years of age. But the performance does vary, even when you get pediatric patients right when you're supposed to. Sometimes the anatomy is just not configured ideally, or the nerve is thin or very small and so can't send a strong signal to the brain. But it is absolutely possible for a child who receives services right away, and stays involved, and gets implanted, to be on a par with their peers by the time they go to school, and then to keep up from then on. It doesn't mean they're not going to need services throughout their life, to help them stay at that level. But they absolutely can."

For James Henry's deaf daughter, the decision to have her own deaf daughter implanted was a relatively easy one. "But a lot of deaf parents would not have made the same decision," Henry said. Indeed, among the deaf, cochlear implants have always been controversial. People who can't hear tend to characterize themselves as either *deaf* or *Deaf*—referring, in the first case, to a sensory fact, and, in the second, to a cultural identification. "The capital-D Deaf tend to feel that cochlear implants are unnecessary—that they are solutions to a problem that doesn't exist," Henry said. "But I think it's a matter of perspective. In my view, a deaf child who gets a cochlear implant has the hearing world opened up to them, whereas members of the Deaf community who remain deaf are isolated from the hearing community. My daughter kind of went against the grain."

Conflicts between the deaf and the Deaf can be fierce. Juliet Corwin, a profoundly deaf fourteen-year-old in Massachusetts, described her own experience in an op-ed piece in the *Washington Post* in 2018. She received cochlear implants when she was a year old, because her parents believed that they would make her life easier. (Corwin's father, Bill, was the president of Clarke Schools for Hearing and Speech from 2007 until 2016, and Juliet and her sister were both students there.) But they also worried that the choice would prevent her from ever being accepted by the Deaf.

"I'm sorry to say that my parents were right," Corwin wrote. "They hired a Deaf ASL [American Sign Language] teacher to work with me when I was only a few months old, but she stopped coming after she found out that I would be getting cochlear implants. When I was a toddler, I was unwelcome in an ASL playgroup. My parents did eventually find a Deaf ASL teacher who respected my family's choice. I've dealt with hearing people not understanding my deafness—staring at the equipment, asking insensitive questions, congratulating me on 'passing' in the hearing world—and I've dealt with Deaf people denying it. . . . I will always feel separated from the hearing world in important ways; I have also had to live with feeling excluded by a community that might have provided assurance that I wasn't alone, that others felt the same way."

I told Henry that I was grateful I hadn't had to make the same decision his daughter did, but that, if I had, I probably would have chosen as she did—and knowing that I was doing what she had done would have made me feel better about my own decision. But the issue turns out to be much more complicated than I had guessed.

Eleven

ASYLUM

My wife and I live in northwestern Connecticut, about an hour west of Hartford. I often drive past the interstate exit for the American School for the Deaf, whose early history I described in chapter eight, and for many years I noticed signs for Asylum Street and Asylum Avenue without ever realizing where those names had come from. (When the school was founded, as the Connecticut Asylum for the Education and Instruction of Deaf and Dumb Persons, the word meant merely "sanctuary" or "refuge.") In the spring of 2018, I spent a day on ASD's campus, observing classes, speaking with teachers and administrators, and learning about deaf education. My visit changed the way I think about deafness, and about my own ears, and about language, and about disabilities of all kinds.

ASD's executive director since 2014 has been Jeffrey Bravin, who was born in Kingston, New York, in 1969. "I'm fourth-generation in a family of deaf individuals," he told me (in American Sign Language, through an interpreter). "My mother and father are deaf, my grand-parents were deaf, on both sides, and my great-grandparents were deaf." Such concentration is rare, and would have stood out even on Martha's Vineyard in 1800. "Ninety-five percent of our students are from hearing families, so for that reason I'm an outsider in the deaf community." He and his wife, who is hearing, have three hearing daughters, "so that chain is broken."

When Bravin was very young, he attended the Lexington School for the Deaf, in Queens—a three-hour drive each way until his family moved to Staten Island to be closer. Lexington had been founded in 1864, and its teaching was entirely "oral": classes were conducted in spoken English, by hearing teachers, and signing was prohibited. "The teacher would talk to you, and you would have to lipread, and we were all expected to understand each other," Bravin continued. "It was exhausting, and we missed a lot." Even hearing people depend to some extent on lipreading, and babies begin to learn to speak in part by studying the mouths of people who are speaking to them (as my granddaughter conspicuously did). But lip-reading is not a substitute for hearing. Even for someone like Gerald Shea, whose remarkable proficiency I described in chapter six, lip-reading provides clues, not a transcript. There are many groups of phonemes that are visually indistinguishable, and if lipreaders don't know in advance at least the context of what's being said, they can quickly get lost. Successful lipreading requires a single speaker, slow and clear enunciation, undivided attention, and, for most people, a significant amount of training. It's useless in group discussions or conversations, and it's nearly useless in deaf-to-deaf communication, and doing it successfully for any length of time is fatiguing.

ASYLUM

In 1979, Bravin appeared in the title role in . . . *And Your Name Is Jonah*, a television movie about a young boy who spends three years in an institution because his profound deafness has been misdiagnosed as a severe mental deficiency. (Sally Struthers and James Woods played his parents.) In the movie, once the mistake has been uncovered, therapists try to teach Jonah to read lips and speak, on the theory that without speech he won't be able to get by in a hearing world. The lessons are unsuccessful, and Jonah's mother realizes that what he really needs is to learn American Sign Language.

Bravin's own deafness was never misdiagnosed, but the movie's plot anticipated significant parts of his future educational experience. Despite the efforts of his teachers at Lexington, he never truly learned to talk. "There were some people who were used to my voice and could understand me, but in general if I'm speaking I just don't have it," he told me. During our interview, he occasionally spoke as he signed, but, even when I had a pretty good idea of what he must be saying, because his interpreter was speaking as well, I couldn't make out even single words. ASL was prohibited at Lexington—a standard rule in oralist schools—but he and his friends signed to one another whenever they could: outdoors, in the bathrooms, under the tables in the lunchroom, and when their teachers weren't watching.

When Bravin was in junior high school, Lexington adopted a different educational approach, called Total Communication. It incorporated sign language, but required teachers and students to sign and speak simultaneously. Most of the teachers used what's known as sign-supported speech, or manually coded English—which, unlike ASL, is not a language but, rather, a method of exactly transcribing English, including its word order and grammar. (For that reason, it's closer to the methodical system that Laurent Clerc learned in France in the early 1800s than it is to ASL.) Bravin told me that the school had some deaf teachers, who used ASL instead. But simultaneously signing in ASL and speaking English is impossible for most people, even if they're fluent in both, because the two languages are conceptually and structurally different. In "Defiantly Deaf," an influential article published in the *New York Times Magazine* in 1994, Andrew Solomon observed that "you can no more speak English while signing in ASL than you can speak English while writing Chinese." Still, Bravin and his classmates no longer had to hide their signing.

In tenth grade, after complaining to his parents that he'd been stuck with the same small group of friends for his entire life—and then, when his parents seemed unsympathetic, purposely failing his classes—Bravin transferred to an ordinary public high school, in Rye Brook, New York. There, he encountered a third theory of deaf education. "At Rye Brook, there were maybe eight or ten deaf students, who were either taught in self-contained classes or were mainstreamed," he said. "I had two adults with me all day, an interpreter and a professional notetaker. The reason for the notetaker was that I couldn't fully pay attention to the interpreter and to the class if I was writing things down. I was able to fully participate, using ASL."

After Rye Brook, Bravin attended Gallaudet University, as both his parents had. While he was there, in 1988, the board of trustees chose the only hearing candidate, from among three finalists, to be Gallaudet's next president, and students shut down the school, in what came to be known as the Deaf President Now protests. Jane Bassett Spilman, the chairman of the board of trustees, raised the anger level considerably by saying that "deaf people are not ready to function in the hearing world"—although she said later that she had been misunderstood by her interpreter. She'd been on the school's board for many years but had never learned ASL. A popular slogan during the protests was "Spilman, learn to sign 'I resign.'" Spilman eventually did resign—and was replaced by Jeffrey Bravin's father, Philip—who six years later was also chosen to be the head of the board of the Lexington School, following similar protests there.

Jeffrey Bravin now presides over a fourth system of deaf education, which is known at ASD as the ASL/English bilingual approach, and is often referred to elsewhere as bilingual/bicultural, or bi/bi. "Our teaching is fully bilingual," he explained—meaning ASL and English. "And there is absolutely no harm in teaching two languages— just as there is no harm in teaching both Spanish and English. We have some children who can both sign and speak—they're fluent in both—and we have some who can't. Every child is different, and we can't predict what an individual child is going to need, so our philosophy is that we offer both."

Earlier that day, I'd sat in on a weekly science lab for a small group of kindergartners, first-graders, and second-graders. All of the children had cochlear implants or hearing aids. Most only signed, among themselves and with their teacher, who was hearing, but a couple of them occasionally spoke. "These kids, among them, actually have a lot of hearing," their teacher told me later. "But they all communicate in the ways they feel comfortable with. Zaire—the one with the high-pitched voice—he's always talking, so I talk back to him, but he's comfortable with sign, and he code-switches all the time, between signing and voicing. A lot of times, the younger kids, if they're from hearing families, they come in talking more, and the kids from deaf families come in signing more. And their preferences change from day to day. Generally, I instruct in ASL, and then if there's a need for reinforcement I use whatever that child's preferred language is. Diana gets English; Katie gets straight ASL. It's very easy to go back and forth between the two, as long as I don't try to do both at the same time."

During the class, most of the students played, as a group, with a mechanical construction game, and they chatted in ASL as they played. LEDs on their implants flashed to show that the devices were charged and working—a feature that's useful with young children but is less likely to be turned on for adults. Every few minutes, an assistant would take one of the kids outside to make a shadow print, in the sun, on a sheet of light-sensitive paper, using flowers and leaves they'd gathered earlier. (They turned the pictures into Father's Day cards.) The five steps involved in making the prints were shown, mostly pictorially, on a whiteboard on one wall—an early-grades version of Thinking Maps, an organizational system used throughout the school. The system is based on flowcharts, tree diagrams, bubble charts, and five other tools for graphically representing multistep tasks. The Thinking Maps program has been adopted by schools for the hearing, too, but an ASD teacher told me that it's especially useful with deaf students, who think visually to begin with. A Thinking Map, because it's multidimensional, bears roughly the same relationship to an ordinarily to-do list that sign language does to English. And the charts' utility extends beyond the classroom: the same teacher, who is hearing, told me that she now uses them at home.

Classes like the one I observed—deaf students in a classroom with other deaf students, being taught by a signing teacher—are not the norm in deaf education today, however. Cochlear implants, improved hearing aids, and other technological innovations have made it possible to give potentially useful levels of hearing to students who, in the past, would have been considered hopelessly deaf, and those innovations, accompanied by a legal and cultural emphasis on mainstreaming students with disabilities, has semi-accidentally created a renaissance of oralism. One consequence is that the composition of ASD's enrollment has changed dramatically. Eighty-two percent of its students now have at least one serious disability, such as autism, in addition to deafness; they're often children whom ordinary public schools were unable to mainstream successfully and sent to ASD because they didn't know what else to do with them. In the early 1980s, ASD responded to that shift by creating a program called Positive Attitudes Concerning Education and Socialization, usually referred to as PACES (pronounced *pay-sees*). It's designed specifically for deaf and hard-of-hearing children "whose emotional or behavioral challenges prevent them from being served in more traditional settings."

Bravin said, "Here at ASD, we're seeing more students at later ages, when the public schools have tried everything, and have exhausted every option, and have realized they can't manage them anymore, and can't teach them—so, 'Let's send them to the deaf school and let the deaf school take care of everything.' But if they come here at twelve, thirteen, or fourteen with limited language, or almost no language, then we have only five or six or seven years to catch them up." Bravin said that it would make more sense to turn that system upside down—to send deaf children to deaf schools first, when they were still very young and at their most receptive to learning language, and then mainstream them later. "Language acquisition happens mainly between birth and eight years old," he said. "That's when they should be here. After eight, if they can talk, if they can hear, then go ahead—mainstream them in public school. Or, if they would prefer to stay here, that's fine, too." The issue is described in ASD's most recent strategic plan: "We find that public schools are keeping children in district from pre-kindergarten until on average, about eighth grade. After they are found to be significantly behind their hearing peers, the decision is made to send the child to ASD. We compare the progress of children who come to us at an early age to their later-enrolled peers and find consistently that children who are exposed to early language interventions and are placed in a communicatively accessible environment do better on standardized tests measuring reading and language acquisition."

Even mainstreamed deaf students who appear to be succeeding in public school—because they do well enough to pass from grade to grade and are therefore never viewed by their teachers as candidates for a different approach—can end up achieving less than they would have if they'd begun with fluency in ASL rather than with a fragmentary grasp of English. A nonintuitive fact about deaf communication is that, for a deaf child, being born to deaf parents is a major advantage, if the parents sign, because such children begin learning sign language exactly when the human brain is wired to begin learning any language, in babyhood. My (hearing) grandson happened to have reached exactly that stage at the time of my visit to ASD. His ability to speak and understand language was noticeably growing daily, if not hourly. He constantly repeated what others said to him, and was beginning to join words together into not-quite-sentences. ("Bird hiding," about a bird, unseen, tweeting in a bush.) He made a pretty good go at "hippopotamus" and "boa constrictor," the two animals with the hardest names in a book he liked a lot, called *The Water Hole*. The thought of a child of his age being mostly or entirely cut off from language is heartbreaking.

Karen Wilson, who is the director of PACES and also ASD's co-ordinator of psychological counseling and evaluation, told me, "You and I, when we were growing up, we learned a lot through osmosis. We heard the radio, we heard the TV in the background, we over-heard conversations, we could hear what other parents were talking about with their children. Deaf kids don't have that. You have to teach them that when they burp they need to say 'Excuse me,' because they don't know that burping makes a sound. 'Really?' 'Yes, it does, FYI.' Ninety percent of deaf children grow up in families that don't sign. If you can't communicate with your child, how do you teach them anything? How do you listen to an adolescent who just broke up with a boyfriend? I mean, imagine that they speak only French, and you don't, and they just came home from school, and they're crying, and you're trying to figure out why."

AT ASD, MY INTERPRETER, like most truly fluent sign-language in-terpreters, was the hearing child of deaf parents—what's known in the deaf world as a CODA, a child of deaf adults. Her hearing is normal, but English is actually her second language: she began signing with her parents before she was physically able to speak. Deaf children who are immersed in sign language as infants have far less trouble, later on, with English—including written English—than deaf chil-dren who didn't learn ASL in infancy and can make out only part of what is being said by their hearing teachers, parents, siblings, and others. And scientific studies have shown that deaf children of deaf parents do better in schools of all kinds than deaf children of hearing parents who don't sign fluently. This seems paradoxical to people who aren't deaf, but, for a deaf child, having hearing parents can be a serious handicap.

Oliver Sacks, in *Seeing Voices*, writes: "Questions of critical age hardly arise with the hearing population, for virtually all the hearing . . . acquire competent speech in the first five years of life. It is a major problem for the deaf, who may be unable to hear, or at least make any sense out of, their parents' voices, and who may also be denied any exposure to Sign. There is evidence, indeed, that those who learn to Sign late (that is, after the age of five) never acquire the effortless fluency and flawless grammar of those who learn it from the start." Hearing parents often disagree, but for many deaf children, and perhaps for almost all of them, the "least restrictive environment" (as mandated by laws like the Individuals with Disabilities Education Act) would be a classroom made up exclusively of deaf students—a classroom like the one I observed at ASD.

All these issues are contentious, not just between the deaf and the hearing but also among people who are deaf. My main guide during my day at ASD was Liz DeRosa, who is the school's director of communications. She has normal hearing, but her husband, an engineer at an aerospace company, is profoundly deaf, as are his brother and sister (a surprising coincidence, since there's no known history of deafness in their family). "I see all the cultural aspects," DeRosa said. "My husband was mainstreamed beginning in first or second grade, and he wears a hearing aid and does not sign. My sister-in-law has a cochlear implant, but she signs fluently, and signing is her preferred mode of communication. My brother-in-law is kind of on the fence. He has a cochlear implant, and he can sign a little bit, and he can speak—he kind of goes with the flow."

DeRosa didn't go to work at ASD because of her husband's deafness; the school was just where the most interesting-sounding job opening happened to be at the moment she was looking. "There's a cultural divide among the deaf, just as there is in my husband's family, and because I work here I can understand both sides," she continued. "My sister-in-law went to the National Technical Institute for the Deaf, at the Rochester Institute of Technology, and she loves signing and she's totally established in capital-D Deaf culture—and I can understand that. I've also come to appreciate just how profound my husband's hearing loss is, and how hard it is for him to navigate every day, because there are people I work with here who can hear me better than he can." She and her husband had just had their first child, who was six months old at the time of my visit. "She's hearing, but I'm taking her to a baby sign class," DeRosa said. "And my husband doesn't want to miss anything, so he's coming, too."

I STUDIED FRENCH FOR SEVEN YEARS, between sixth and twelfth grades, and topped myself off with a bonus year, when I was a junior in college, in order to meet an English department graduation requirement. In other words, my French education began exactly when it might have concluded, after my language-acquisition window had presumably closed, and my efforts dragged on futilely through most of a decade of vocabulary lists and pop quizzes and conjugations of irregular verbs. Some people have no problem learning languages later in life; my wife's brother didn't begin to study Russian until after college, yet he became so fluent that he once served as a simultaneous translator for Andrei Sakharov. But the gene that makes that possible is apparently missing from my side of the family. My father, who also studied French in high school, once spoke to a Frenchman sitting at the next table in a restaurant in Paris, on a trip with my mother, and the Frenchman spoke back—but neither man was able to understand what the other was saying. At last, my mother identified the problem: the Frenchman didn't realize that my father was speaking French, and my father didn't realize that the Frenchman was speaking English.

I now wish that I'd started my second-language education years earlier, before I could walk and talk, and that I'd studied ASL instead of French. If we all knew how to sign, as the residents of Chilmark did in the seventeenth, eighteenth, and nineteenth centuries, some of the biggest benefits would go to those of us who hear perfectly well, or used to hear perfectly well. Just think: you'd no longer dread going to sports bars, and you'd be able to converse with people at parties without screaming into their ear, and no one would shout at you from across the kitchen because they couldn't hear you over the TV, and you'd be able to converse with toddlers who couldn't talk yet—like my grandchildren, who learned a dozen signs in daycare and kept using them for a while after they'd begun talking, because signing is useful and efficient. For anyone who hangs around with the sorts of people that I increasingly hang around with, there would be significantly less reliance on *"Huh?"*

Even if we don't learn to sign, we can embrace the example of the Martha's Vineyard deaf community, and realize that what we think of as handicaps are often cultural constructs created out of ignorance.

Those olden-days farmers and fishermen had no access to cochlear implants or Bluetooth-enabled hearing aids, and, because they didn't travel to the mainland very often, they seldom had to deal with people with whom they didn't share a language. But in the most important ways their world wasn't radically different from ours: most of them could hear, but some of them could not. The main difference is that they worked things out in ways that most of the rest of us never have. Technology can help us bridge the divide between those who can hear and those who can't, but we need patience and sympathy and understanding as well.

Twelve

THE MICE IN THE TANK

One day in 2017, at Harvard Medical School, David Corey and his colleague Bence György showed me a sequence of three videos on a monitor hanging on the wall of Corey's office. (Corey is the professor who also showed me an electron micrograph of a mouse's hair cell, described in chapter three.) In each of the videos, a mouse was dropped into a tank of water. The first mouse paddled back and forth, trying to escape. "This is a normal mouse, and that's the way a normal mouse swims," Corey said. "He knows which way is up, and he always keeps his head above the water." The second mouse had been bred with a specific genetic mutation, as a consequence of which it could hear nothing and had no sense of balance. It thrashed wildly underwater, as though caught in a turbulent current. "He doesn't know which way is up, and he just tumbles, and we have to rescue him," Corey said. The third mouse had the same mutation, but had been given a functioning version of the faulty gene. "He's not quite as good a swimmer as the control mouse, but he has enough of a balance system now to keep his head above the water," Corey said. The treated mouse was also able to hear, as it demonstrated to the researchers by responding to a loud hand clap.

The transformation of the treated mouse seemed miraculous to me—and it was—but Corey and György both cautioned me not to draw unwarranted conclusions. Only about one child in a thousand is born with a genetic hearing loss, and such losses can be caused by defects in any of more than a hundred different genes. In order for treatments to succeed, the responsible genes have to be identified, and the interventions have to be tailored specifically to them. And, even if a successful treatment can be found, the number of individuals who might benefit from it will necessarily be small, meaning that the cost of treatment will necessarily be high. Nevertheless, Corey said, hearing loss caused by genetic mutations will probably turn out to be easier to solve than hearing loss caused by exposure to noise or ototoxic substances. With genetic losses, researchers have to do the equivalent of turning on or off a single malfunctioning switch, while with acquired hearing loss they have to figure out how to rebuild some of the most complicated micro-infrastructure in the body.

"There may be treatments for hereditary hearing loss before acquired hearing loss," he continued. "I would say that within five years, and certainly within ten, you're going to see gene therapy for certain forms of hearing loss—although whether they're going to work or not I can't promise." One challenge, although it's by no means the only one, is that therapies that work in mice won't necessarily work in humans. Corey said that, in his program, what they were hoping was that by 2022 or so they will be ready to do "preclinical testing," probably in primates, to be followed a few years later by tests with people.

In the mouse in the third video he showed me, the functioning replacement gene had been delivered to its cochlea by a "viral vector"—a harmless virus that transported the new gene to the site without causing mischief elsewhere in the mouse's body. "I like to say it's like a Trojan Horse, but instead of soldiers it's filled with doctors," Corey said. György added, "It's not going to be the solution for every type of genetic hearing loss, but for some of them it could definitely restore hearing." Viral vectors, they both said, may also be useful in treatments for nongenetic hearing loss. So may bisphosphonates, which are drugs that are used in the treatment of osteoporosis. They work on osteoporosis by avidly binding with bone, and, because the cochlea is surrounded by bone, researchers hope that they can be used to securely implant therapeutic substances inside it, like microscopic medicine dispensers.

THE INAUGURAL BREAKTHROUGH in the field of hearing restoration occurred in the late 1980s. Edwin Rubel, who at the time was a member of the faculty of the University of Virginia and is now at the University of Washington, was interested in establishing a "timeline" of the effects that ototoxic drugs have on inner ears. "We were looking for the first sign of damage, so that we could start to figure out the biology of hair-cell death," he told me. At his direction, a surgical resident in his lab administered ototoxic drugs to young chickens, then waited varying lengths of time before euthanizing them and examining their cochleas. What the resident found was that devastating hair-cell dam-age occurred very quickly after the chicks were given the drugs—but that in chicks euthanized a couple of weeks later the damage appeared to be less severe. Rubel told him he must be doing something wrong, and sent him back to the lab. The resident repeated the experiment, twice, with the same results. Rubel decided that he'd better try it himself.

"And there was less hair-cell death," he told me. "So what was hap-pening? Either we were merely injuring the cells and they were re-covering, or we were destroying the cells and they were regenerating." No one at that time believed that hair cells could regenerate, but Rubel and his colleague were able to show that that was indeed what was happening. And, essentially simultaneously, Douglas Cotanche (who was then at the University of Pennsylvania and is now at Boston University) discovered the same thing, also accidentally, in chicks that he had deafened with noise. Researchers elsewhere eventually proved that the same kind of regeneration occurs in other animals, too, including fish. Rubel said, "What we found is that every verte-brate regenerates hair cells—except mammals."

I asked why mammals were the exception.

"It's very clear that there was selection pressure," he said. "And regeneration was probably selected out, in my opinion, with the evolution of high-frequency hearing—which early mammals really depended on." Early mammals were small and nocturnal, and to avoid being eaten they had to localize predators and precisely distin-guish threats from non-threats. Most organs, he said, can afford to lose and create cells, but high-level hearing is different. "In the ear, any new cell will change the frequency organization," he said. "So we don't want new cells. We want to keep this thing stable." He went on, "In order to develop the mechanics to hear and process very high-frequency sounds, the entire mammalian hearing system, from the middle ear through the cochlea and even into the brain, went through huge transformations—because the mechanics of the system are so delicate." To preserve stability and conserve limited resources, selection favored durability and consistency over repair, and mammals evolved inhibitors that kept potentially disruptive new hair cells from forming.

In 2011, Rubel was instrumental in creating the Hearing Restoration Project, a consortium of scientists who agreed to work together to find ways to reverse deafness in humans, partly with funding from the Hearing Health Foundation, the New York–based organization I described in chapter six. Rubel told me, "In 2013, Albert Edge, a member of the consortium, led a group of scientists who showed that very young mice with noise-damaged ears could recover some hearing if a drug was delivered directly into their inner ears shortly after they were deafened." It was the first time that mammals had proved able to regenerate hair cells. The drug suppresses the activity of a protein that prevents hair cells from being created by so-called supporting cells, which are cells in the cochlea that function something like stem cells. Rubel said, "What that shows, beautifully, is that there is something there that can support regeneration. We just have to figure out how to goose it along."

Edge is the director of the Tillotson Cell Biology Unit of the Eaton-Peabody Laboratories, at Mass. Eye and Ear. Eaton-Peabody was founded in 1958 and is the largest hearing research institution in the world. I visited twice, in 2017 and 2018. If you're traveling on the

Red Line, you get off at the stop closest to the Boston side of the Charles River, which is the stop for both Mass. Eye and Ear and Massachusetts General Hospital, and, if you're early for an appointment at either place and don't mind doing this sort of thing, you can hang out in the lobby of the Liberty Hotel, which is right between them. You can sit on a comfortable couch with a cup of coffee from the big urn over by the stairs and read the newspaper on your phone. Nice restrooms, too.

During my first visit to Eaton-Peabody, Niliksha Gunewardene, a postdoctoral fellow, took me up a flight of stairs to a small room containing a piece of equipment about the size of a washing machine. "This is the chamber we use to deliver high levels of noise, to kill off hair cells," she said. On a black-and-white video monitor I could see that the chamber contained a small cage with several four-week-old mice inside. The mice appeared be running around normally, but they were being subjected to two hours of steady noise at above 100 dB—enough to ruin their hearing. "After exposure, we check their auditory function, to make sure it's been lost," she continued. "Then we do surgery, to deliver drugs to their cochlea through the round window, and then we test auditory function again, at one week, one month, or three months post-surgery." She told me that she and her colleagues were currently able to improve the hearing of a deafened mouse by about 15 dB. "Which is good, but we'd like to improve it further," she said.

Not long before my visit, Edge and several other researchers had succeeded in causing supporting cells they'd extracted from normal mice to divide and differentiate into large clusters of hair cells. Danielle Lenz, a coauthor of the paper describing that experiment, put on latex gloves, washed her gloved hands with alcohol, and removed two plastic trays from a shelf in an incubator, then placed them on a microscope. She said, "In the second tray you can clearly see the organoids that have been formed from the single cells in the first tray, and you can see that they are multicellular." A normal mouse cochlea contains about three thousand hair cells; Lenz and her colleagues had been able to produce as many as a quarter of a million in a single petri dish in just under three weeks. On the computer on her desk she showed me images of some of the hair cells they'd grown in vitro. "Here you can see the stereocilia," she said, pointing to the screen. "And here as well."

The benefit, for now, is in the laboratory. "These cells are not quite there yet for implant, but they are a great research tool," she said. Having access to an essentially unlimited supply of living hair cells makes screening drugs and other potential remedies easier, and researchers can minutely study the steps in the transformation of supporting cells into hair cells. And there are hopes for bigger things. Edge told me, "The ear is maybe a little bit behind the eye, in terms of treatment, but there has been a lot of progress, and between the soldiers and the baby boomers there's a lot of interest."

THE CHIEF OF OTOLARYNGOLOGY at Mass. Eye and Ear is D. Bradley Welling. He grew up in a Mormon family in Salt Lake City, and, after his freshman year in college, he spent two years as a missionary in Japan. He earned his MD at the University of Utah and got a PhD in pathobiology at Ohio State. The experience that persuaded him to choose hearing as his medical specialty, he told me, was watching a stapedectomy—the operation I described in chapter six that instantly restores hearing in most people who have otosclerosis. He is a clinician and a scientist as well as the head of a large department. His longtime research focus has been on a hearing problem that is far more exasperating than otosclerosis: neurofibromatosis type 2, commonly called NF2. It's a rare genetic disorder characterized by the proliferation of tumors similar to the one that ravaged Stephen Colbert's right ear, but in NF2 the tumors arise on both sides of the head and, often, well beyond the ear itself. NF2 causes hearing loss, tinnitus, balance problems, impaired vision, and facial-nerve paralysis, among other horrors. It can destroy the cochlea and the auditory nerves connected to it, and when that happens hearing aids and cochlear implants don't help. Some patients are given a device called an auditory brain stem implant. They then wear a microphone behind their ear, and the microphone sends signals to a chip placed under the skin, and the chip sends signals to electrodes inserted directly into auditory centers in the brain. A description on the website of the Mayo Clinic says that an implant "helps most people distinguish sounds such as telephone rings and car horns," while providing "word recognition" and "general sound cues" to some. "It's a miserable disease," Welling said.

Welling himself has less than perfect hearing. He has trouble in his department's boardroom, which contains a fan that whirs in the frequency range of human speech. "If I happen to be sitting on that side of the room, I put on my Bose Hearphones and crank them up," he said. He has tinnitus as well. I told him that there was an ambient hum in his office (coming from the HVAC system) whose frequency serendipitously made it a perfect masker for my own tinnitus—but he couldn't hear the hum. He said, "There's a fan over here that I can hear. Can you hear that one?" I said I could.

Welling and I are about the same age—we're both in our mid-sixties—and we had similar adolescent experiences with sound. "When I was about fourteen years old, my friend Danny and I used to sit in his father's sound room, in their basement," he said. "It was about half the size of this room, and he had four Klipschorn speakers—these big theater speakers—one in each corner." The boys would sit on a couch in the middle of the room and crank up Led Zeppelin or Pink Floyd. "It sounded like a flying saucer was landing on top of you. Danny would give us a Styrofoam hat to hold, and that hat would just shake like crazy. You could feel your *clothes* shaking." Kids nowadays who listen to loud music through earbuds have nothing on Welling's and my generation. "When we'd come out of it, after an hour or two, our ears would be ringing, and we wouldn't be able to hear very well for twenty-four or forty-eight hours," he continued. "And that's actually a lead-in to something I want to talk to you about."

Symptoms of the type experienced by Welling and his friend used to be considered temporary: you listened to something really loud and your ears felt broken for a day or two, but then they got better. "But we now know that hearing loss in those scenarios is not reversible, and that the damage is permanent," Welling said. "And the researchers who established that, and delineated what the underlying pathology is, are right here at Mass. Eye and Ear."

FOR DECADES, the conventional wisdom was that ears could recover from even very loud sounds if the exposure didn't last for a long time.

And that belief was supported by the results of standard hearing tests, like the one I was given at Starkey. If Welling and his friend had been tested both a week before and a week after a single visit to the basement sound room, their audiograms might very well have shown no change. And there was pathological evidence that ears could recover from even very loud sounds. The super-magnified computer image that David Corey showed me of mouse stereocilia thrown around like tree trunks was from a study that was actually concerned mainly with the ability of hair cells to return to normal after a period of intense exposure. If the same stereocilia bundle had been imaged a few days later, Corey told me, everything might have looked fine.

But for years there had also been suggestions that something more complicated was going on, and that ears could be harmed in ways that standard testing didn't detect. Clinicians had known, for example, that two people with identical audiograms could have markedly different abilities to comprehend speech, especially against a background of noise, as in a classroom or a busy office. What no one understood was why such differences existed. Many audiologists concluded that the difficulties some patients reported with comprehension must be mainly psychological, and that, for example, a military veteran who did well on hearing tests but had trouble talking on the telephone must be suffering from something like post-traumatic stress disorder. But scientists now know that the conventional wisdom is wrong and that, although ears can indeed recover fully from some intensities and durations of exposure to loud sounds, permanent damage occurs more quickly, and at lower decibel levels, than had previously been thought.

One of the people who solved the riddle was Sharon Kujawa, the director of audiology research at Mass. Eye and Ear. She began her professional career as an audiologist, and saw patients in a clinic, but then went back to graduate school. In the mid-1990s, she won a postdoctoral fellowship at Mass. Eye and Ear, where she worked with Charles Liberman, the director of the Eaton-Peabody Laboratories. When the fellowship ended, she joined the faculty at the University of Washington, and there she took part in a retrospective investigation of hearing loss among subjects of the Framingham Heart Study—a longitudinal medical research project commissioned by Congress in 1948 and still under way.

"I was working with George Gates, an otologist who's now retired," Kujawa said. "He was trying to get at this question: If you are noise-exposed at some time in your life, and then the noise stops, does that change the way your ears age, going forward?" A widely held assumption at the time, she said, was that hearing loss caused by noise was not progressive, and that, if a dangerous exposure ceased, any damage that had resulted from it would stabilize. Gates and Kujawa identified older male participants in the Framingham cohort who had worked at loud jobs and, during their employment, had taken hearing tests whose results suggested that the jobs had caused damage to their ears. Then the scientists compared those audiograms with audiograms taken much later, after the men had presumably retired, to see if their losses had become more severe over time.

"And, when we looked at the data, that's what we saw—that the losses got worse, and that ears and hearing had aged differently after exposure to noise," Kujawa continued. She and Gates couldn't be certain that the continuing losses hadn't had other causes—the retirees could have taken up noisy hobbies, for example—but then Kujawa thought of a way to test the question directly, by studying lab mice from a previous experiment, in which she, her postdoc adviser Charles Liberman, and Bruce Temple, a molecular geneticist, had investigated inherited resistance to noise damage.

"I had taken some of the mice from that experiment and put them back in the animal-care facility, and they had been there for as long as two years after receiving a single two-hour exposure to noise," she continued. "I'd also saved some unexposed mice—their cage mates— as controls." Kujawa and Liberman got a grant from the National Institutes of Health, tested the hearing of all the mice, and found the same effects that Kujawa and Gates had identified with their Framingham study: "The ears of the mice that had had a single prior noise exposure in their youth had aged differently from the ears of the other mice," she said. She and Liberman then euthanized the mice and dissected their cochleas—and, when they did that, they found that the hair cells were undamaged, suggesting that the mice's hearing had recovered fully from the noise they'd been exposed to. But, farther along the auditory path, between the mice's hair cells and their brains, they found something that shocked them: the noise-exposed mice had suffered "massive neural degeneration." Their hair cells looked fine, in other words, but the nerve fibers were shot.

Kujawa and Liberman published their results in 2006 but couldn't fully explain what they'd observed. They did that three years later, in another paper, in which they were able to pinpoint what was, in effect, ground zero of the progressive hearing loss they'd found: the synapses that connect hair cells to auditory nerve fibers—neural terminals that function like electrical sockets.

No one, previously, had thought much about damaged synapses. The main reason is that they're difficult to observe, because, in addition to being extremely small, they're surrounded by other cellular structures, and can be resolved only by using a particularly tricky version of a protein-detecting technique known as "immunostaining." Liberman and Kujawa succeeded by using fluorescent molecules that bound themselves to two specific protein types, one inside the base of the hair cells and one on the other side of the neural connection. When they did that, the proteins lit up like Christmas lights, red on one side and green on the other. "Now we could see everything in a light microscope," Liberman told me. He and Kujawa were able to determine exactly which connections had become "unplugged," and they were able to construct a dramatically revised understanding of how noise causes hearing loss in mice. By extension, they were also able to offer a likely explanation of why most of us humans, as we get older, increasingly complain about having trouble understanding what other people are saying to us, even if our audiologists tell us we're basically O.K.

David Corey told me, "What happens is that, if you start driving the hair cells too much, these nerve fibers that are picking up the information withdraw, and detach from the hair cell, and now the hair cell can release its signal all it wants, but the nerves aren't listening anymore." Worst of all, detachment can occur after exposures that had always been believed to be harmless. Stéphane Maison, a researcher at Mass. Eye and Ear who works with both Kujawa and Liberman, told me, "Before 2009, we thought that if, for instance, you go to a club or you go to a concert and you're exposed to a lot of sounds, you might hear a ringing, a buzzing in your ears, from tinnitus, and you might feel a fullness, like cotton in your ears—but, if you are lucky, the next day you wake up and everything seems to be fine. Then you have a hearing test, and your audiogram hasn't changed, so you're good. That's what we thought. But in animal models in 2009 we discovered that that was not the case."

If sound exposures of this type cause permanent damage to the inner ear, why do people who have experienced them often seem unaffected when their hearing is tested? The reason is that the ability to detect discrete pure tones—the kinds of tones an audiologist plays for you in a sound-isolated booth when you finally make that appointment to find out whether you need hearing aids—doesn't require a fully intact auditory apparatus. "Literally, you can lose about eighty percent of the synaptic connections before that loss shows up on an audiogram," Liberman told me. What does diminish immediately is the ability to make sense of complex sounds, and especially the ability to understand speech against a background of noise.

Kujawa and Liberman called the damage that they discovered "cochlear synaptopathy." A scientist at a different institution, a few years later, called it "hidden hearing loss"—and that name has stuck. It's an evocative term, and it's not a tongue-twister, but it's a misnomer because it suggests that the condition is something distinct and mysterious; a more accurate term, in most cases, might be just "hearing loss." Indeed, it now seems possible that nearly all sensorineural hearing problems at least begin with damage to the synapses, and that such damage is "hidden" only in the sense that, until fairly recently, no one had tools that were capable of seeing it. Maison told me, "People who wake up one day with sudden single-sided deafness [SSD] usually get their hearing back after treatment, but when they do they often say that something is not the same—that their discrimination is off. Ménière's patients, too. When the Ménière's resolves, they no longer have vertigo, but they have tinnitus and very poor discrimination. All these things—SSD, Ménière's, tinnitus, aging, noise—they seem to have this one thing in common. No matter what the insult is, there is damage to the nerve."

IN ONE WAY, all of this seems like horrible news: if Kujawa, Liberman, and the others are right—and it now seems likely that they are, since the same effect has been found in every animal species that's been tested, and in preliminary tests with humans—our ears can be damaged by sound levels that have long been thought to be harmless, and the steps we typically take to protect ourselves, when we bother to take them, are therefore insufficient. But in another way it's very good news. The reason is that repairing or reattaching broken synaptic connections seems tantalizingly achievable. David Corey told me, "When you look at these damaged synapses with microscopy, you see that the close attachment between the nerve fiber and the hair cell is separated as little as one ten-thousandth of an inch." That's enough of a gap to block the signal, but it's so small that finding a way to bridge it seems possible. "Basically, if you could coax the nerve fibers back, you might be able to repair that damage," Corey continued. "So if you could get the hair cell to release some kind of factor that would say to the nerve fiber, 'Come to me,' then you might get reconnection." Indeed, Liberman and two colleagues have successfully reconnected the synapses of a deafened mouse twenty-four hours after noise exposure, by delivering certain naturally occurring substances into the cochlea by way of the round window. "We showed that these treated mice had near-complete functional recovery, suggesting that the regenerated synapses were functional," Liberman told me. He and Albert Edge are among the founders and technical advisers of Decibel Therapeutics, a Boston-based company that hopes to turn some kinds of hearing loss and tinnitus into transitory conditions, reversible with drugs.

Regenerating damaged hair cells poses a bigger challenge. Corey said, "One possibility would be to replace a dead hair cell with some kind of stem cell. But it's really difficult to get a stem cell to migrate into this part of the cochlea and somehow fit itself into that structure exactly where it's supposed to go, and then perform exactly the way it's supposed to perform." One difficulty is that hair cells aren't free-floating sound transducers; they're single components in an intricate hearing machine, in which all the parts have to fit together just so. Inducing a cochlea to randomly grow new hair cells would be about as useful as inducing a piano to randomly grow new strings. Still, there has been encouraging progress in this area as well. In 2019, I visited the headquarters of Frequency Therapeutics, a startup in Woburn, Massachusetts, a dozen miles north of Boston. The company was then testing a drug therapy that, in a preliminary human trial, had made what it described on its website as "a statistically and clinically meaningful improvement in key measures of hearing loss, including clarity of sound and word understanding." The drug they're using, called FX-322, is intended to reverse hearing loss "by regenerating hair cells through activation of progenitor cells already present in the cochlea"—the holy grail. Phase 2a trials began in late 2019. If everything works out, the drug will be "administered locally by an ear, nose, and throat specialist, using a standard in-office procedure." Although tinnitus was not a focus of the initial study, some participants whose hearing scores improved reported that the ringing in their ears was reduced. The company's researchers believe that FX-322 may also be effective in treating certain other degenerative diseases, among them multiple sclerosis.

Many other efforts at restoration are under way, at companies and research centers all over the country. On the West Coast, Edwin Rubel, a co-discoverer of hair-cell regeneration in chickens, is working on drugs that could be used to prevent sensorineural hearing loss from occurring in the first place. He's currently experimenting not with chickens but with zebra fish, which are easier to study because their auditory organs are on the outside of their bodies. If he and his colleagues are successful, drugs they develop could be given to patients before they're treated with ototoxic medications, including chemotherapy drugs and antibiotics like streptomycin. He told me that he believes that such drugs will be available before hearing-restoration treatments are, but he's optimistic about the entire field—which, as a founder of the Hearing Restoration Project, he was instrumental in creating. He said, "I have a friend who studies spinal-cord regeneration. He told me, 'You guys really have an advantage; you know which cells you've got to replace.'"

Thirteen

VOLUME CONTROL

S haron Kujawa and Charles Liberman's work on hidden hearing loss suggests, among many other things, that American noise-related workplace protections are inadequate. In 1970, Congress passed the Williams-Steiger Occupational Safety and Health Act, a bipartisan bill whose purpose was to eliminate hazardous conditions in many American workplaces, and President Nixon signed it. The law led to the establishment, the following year, of the Occupational Safety and Health Administration (OSHA), which issues and enforces regulations related to workplace safety, and the National Institute for Occupational Safety and Health (NIOSH), which supports those regulations by performing research and since 1973 has been part of the Centers for Disease Control and Prevention. Kujawa told me, "Both OSHA and NIOSH are meant to protect workers over a forty-year occupational life, and that means workers who are exposed to high levels of noise eight hours a day, five days a week, for forty years. Yet the damage we've identified in animal models occurs after discrete exposures. I can't even imagine what we're doing to workers' ears with what we allow."

OSHA's guidelines are complicated, and they involve formulas and averages and a 5-decibel margin of error. Basically, though, they say that if you work in a covered industry you can legally be exposed to eight continuous hours of 90-decibel noise (motorcycle eight meters away, lawn mower), or to two hours of 100-decibel noise (New York City subway car, jackhammer, kitchen blender, snowmobile), or to thirty minutes of 110-decibel noise (car horn one meter away, chain saw)—every day of your career. You can also be exposed, occasionally, to "impulse" noises as loud as gunshots. Employees whose exposure is close to the maximums are supposed to be tested at least once a year. Probably the best that can be said about the rules is that they're better than nothing. Still, the mandated hearing tests are just the standard ones, which Kujawa and Liberman's research has suggested can't detect synaptic damage until the vast majority of the outer hair cells have been destroyed. A retired electrical engineer who worked on cochlear implants at 3M in the early 1980s (and has been meticulous about protecting his ears ever since) told me that he thinks the maximum safe sound level for sustained exposure is probably closer to 60 decibels.

Compounding the risk is the fact that enforcement of even existing standards has always been inconsistent. Robert Dobie told me, "OSHA visits less than one percent of workplaces, on an annual basis, and they do that almost entirely in response to complaints. So if you run a noisy factory and you ignore noise regulations, you know that nothing is going to happen to you unless someone complains." Employers who are subject to OSHA regulations, and even their employees, often argue that the hearing-protection standards the law requires them to meet are unnecessarily strict and too costly to implement. But every cochlear synaptopathy study conducted so far suggests that they're not strict enough, in terms of actually preventing lifelong hearing loss, and that workers in the noisier sectors of the economy are permanently damaging their ears. Liberman told me, "The workplace guidelines that have evolved over the past thirty or forty years are all based on the premise that, if an audiogram returns to normal after an exposure, then the exposure was really a nonexposure. And I think that everyone in the field would agree that that just isn't true."

At even greater risk are people whose jobs aren't subject to OSHA's requirements. "These are people in construction and agriculture and oil-and-gas drilling—industries that have a lot of mobile and transient workers, and are therefore exempt to varying degrees," Dobie said. "There are at least a million people like that. It's harder to develop a hearing-safety program for them than it is for people who work in a factory, but it should be done, because they're really not being protected at all." I see such people almost every day: landscapers using lawn mowers and chain saws, town maintenance crews operating heavy machinery, carpenters working with power tools, me vacuuming the rug in my office. In 2020, I was interviewed for an informational video about hearing. During a break, all the sound engineers came out of the control booth, as a group, to ask me questions about hearing protection, because all of them, after decades of daily headphone use, had suffered what they described as noticeable losses. Also at risk—according to studies I've read, conversations I've had with sufferers, and emails I've received from readers—are airline pilots, radio-show hosts, soccer fans in countries outside the United States, members of college marching bands, and people who work in kennels, veterinary offices, animal hospitals, and animal shelters. I once played golf on a course that was across a broad highway from a stock-car racetrack. A single car was doing practice laps—a single car—yet the engine noise was so loud that on some holes my playing partners and I had trouble conversing. The Seattle Seahawks and (my beloved hometown) Kansas City Chiefs are said to have the loudest fans in the National Football League—a point of local pride but also a health hazard. Peak sound levels above 140 decibels have been recorded in both stadiums. That's right up near gunshot on the Owen scale, and plenty loud enough to cause permanent hearing damage.

The only plausible remedy for the foreseeable future is for people who are routinely exposed to dangerous noise levels—that is, virtually all of us—to take responsibility for our own ears, and for employers of all kinds, including those in unregulated industries, to decide that truly protecting the hearing of the people who work for them is the right thing to do, if only because it's in everyone's economic self-interest to prevent workers from deafening themselves. In 2011, I traveled to Bogotá, Colombia, on a reporting assignment, and allowed a man there to shoot me in the stomach from a few inches away with a .38-caliber revolver. (His business was making fashionable bulletproof clothing, and I was wearing one of his jackets.) Before he shot me, he put on earplugs and made me put on earplugs, and he shouted across his company's main manufacturing area to warn the several dozen workers there—most of them women sitting at sewing machines—to do the same. Those workers complied without looking up from what they were doing. (He demonstrates his products pretty often.) At the time, his caution struck me as excessive, since he was going to shoot me just once, and only with a pistol. But now I'm glad that I wore the plugs.

If the owner of an unregulated business in a country that until recently was known mainly as the world's leading exporter of cocaine can voluntarily supply hearing protection to all his employees, then the owners of unregulated businesses in other countries, including ours, can do the same. Deafness is expensive. Earplugs aren't.

I WAS BORN IN 1955. The relationship between noise and hearing loss had been definitely established by then, but my friends and I didn't think much about it, if we thought about it at all. We had been warned that slingshots and BB guns and darts and arrows and gym towels could blind us if we aimed them at each other's faces, but I have no memory of being similarly advised about the dangers of sound. A toy store that my friends and I often rode our bikes to sold lots of dangerous things, including many that can't possibly have been good for our ears: Greenie Stik-M-Caps (for the hard plastic projectiles fired by my Mattel Fanner 50); red roll caps (for hitting with hammers—ideally, one full roll at a time, producing flame, a plume of acrid smoke, and an ear-stabbing explosion); and model glue that still smelled like model glue (because it hadn't been reformulated yet to deter sniffing). The aural threat posed by the glue was that we used it to build model cars and airplanes that we then blew up with firecrackers, and if we had any firecrackers left we threw them at each other. At every summer camp that I attended, from third grade on, I fired .22-caliber rifles at paper targets without wearing (or being offered) ear protection. In junior high school, I listened to Steppenwolf's "Born to Be Wild" while lying on the floor of my bedroom with loudspeakers leaning against my head, and I bought a pair of headphones so that I could crank my stereo when my parents were at home. My friends and I went to concerts by Jefferson Airplane, the Grateful Dead, Led Zeppelin, Canned Heat, Janis Joplin, the Rolling Stones, and the Who, among many others (typical ticket price: $3.50, except for the Stones, who charged $5.00), and we always sat as close as possible to the banks of loudspeakers at the front of the stage, so that we would receive the full effect.

My friend Duncan's family had a white 1963 Chevrolet station wagon, known as the Economy Wagon. Its odometer had frozen at more than a hundred thousand miles, and its steering wheel pulled hard to the left, so that to go straight you had to turn it hard to the right, and the passenger side of its front seat had become unmoored, so that it slid forward in its track when you slowed down, and slid backward when you accelerated again—a half-scissoring motion that a driver had to anticipate and compensate for, by adjusting pedal pressure. One summer when we were in high school, my friends and I rolled down the Economy Wagon's backseat windows and filled the openings with rectangles of paperboard, in the center of each of which we'd punched a hole just big enough for a bottle rocket. A classmate of ours had a newer Chevrolet, a sedan, and we modified it in the same way, then drove around Kansas City for a couple of hours shooting bottle rockets at each other and trying to throw firecrackers and smoke bombs into the other car's open front-seat windows. At some point, the guys in the other car managed to throw a cherry bomb into the

Economy Wagon, and the explosion was so loud, inside that enclosed space, that for a few seconds my mind went utterly blank. Then we went back to racing around and launching explosives at each other.

In college, I spent a summer cutting grass at an apartment complex in Colorado Springs, and it never occurred to me or my fellow lawn-crew members to use ear protection, even though we ran our lawn mowers all day. My wife and I slept with earplugs when we lived in Manhattan, but I didn't wear them when I rode the subway, walked along Second Avenue, or paused to watched fire engines and police cars inch their way past drivers frozen in Midtown gridlock, their sirens and horns at maximum volume. In 1985, my wife and I and our infant daughter moved into an old house in a small town in northwestern Connecticut. Nights were so quiet that I could hear owls and coyotes in the woods across the road, but I also became an avid home-improver and gradually acquired an extensive collection of noisy power tools. My wife had begun worrying about her own ears years before—one of her grandfathers was at least as deaf as my grandmother—and kept a big jar of foam earplugs in the kitchen and wore them even while using her food processor. I scoffed. She bought me a pair of headphone-like ear protectors to wear while I worked on the house or mowed the lawn, but I seldom used them unless I was working in a place where she could see me.

One summer, with two friends who also owned houses that needed lots of attention, I signed up for a woodworking class at a craft center in another town. The class was devoted to the router, a power tool that is basically a high-speed electric motor with handles, plus a chuck to which you can attach an amazing variety of murderously sharp cutting bits. In order to be admitted to the class, my friends and I had to show that we'd brought eye and ear protection, and our instructor, a professional cabinetmaker, spent the first half hour discussing workshop safety. He told us that we should never use our routers or any other power tools without first putting on earplugs and safety goggles, and he showed us the particular type of earplug that he said we ought to buy. He told us horror stories about people he knew who had been blinded by flying wood chips or had fed their fingers into their router bits, and he said that woodworkers were as prone to hearing problems as aging rock stars.

At last, our instructor was ready to demonstrate a few basic cuts. Before he turned on his router, he did not put on his safety goggles and he did not put on his earplugs. To get a better view of what he was doing, he bent down close to the table and placed his face a few inches from the spinning bit, which was emitting a high-velocity blizzard of woodchips. To keep the chips out of his eyes, he squinted, turning his eyelashes into protective screens. Even from where I was sitting, halfway across the room, the whine of his router was unpleasantly loud and high-pitched. Still, I didn't put on my ear protection. Like everyone else in the room, I had placed my safety equipment neatly on the desk in front of me. Also like everyone else, I got out of my chair and moved closer to the front of the room to get a better look.

I don't think our instructor felt hypocritical for failing to practice the self-protective measures he had just spent such a long time preaching. In our classroom, as in many woodworking shops and job sites, safety was an idea that we paid homage to rather than a set of procedures that we actually followed. We invoked the idea of safety, like a charm, before we went to work. But actually wear earplugs? Forget it.

Carpenters and woodworkers nowadays are more likely to wear hearing protection than they were then, but many still don't, even if they've already lost a significant amount of hearing. Not long ago, I watched a *This Old House* video about how to install a kitchen exhaust fan. Tom Silva, the show's general contractor, used many power tools in that project, including saws and drills, but neither he nor the woman assisting him appeared to be wearing anything in or over their ears. And that's the reason that, when you talk to people you've hired to remodel your house or repair your appliances or cut the grass in your yard, you often have to speak up.

IF I COULD RELIVE MY LIFE, I'd wear ear protection through my entire power-tool period, as well as during much of my adolescence. I can't do that, of course, but in recent years I've become much more cautious, in the hope of preserving as much of my remaining hearing as I can. Recently, I wore those ear protectors my wife gave me, the ones I used to scorn, while doing nothing more cacophonous than pounding in a few big nails. Why take chances?

On the advice of James Henry, the VA research scientist and tinnitus expert, I now own several sets of so-called musician's earplugs. They're good for musicians because they reduce the overall level of sound but maintain almost the full sonic spectrum—unlike regular foam earplugs, which disproportionately mute high frequencies. The ones that Henry wears cost several hundred dollars and were custom-made from ear-canal impressions taken by an audiologist; mine are off-the-shelf, from Amazon, and cost $10 or $15. Mine are made by Etymotic Research, which has been making high-fidelity headphones and other audio devices since the early 1980s; it sells a full range of musician's plugs, including not just the cheap ones that I bought but also fancy ones like Henry's, and it makes the filters used by a number of other manufacturers. In 2018, Etymotic merged with Lucid Audio, a company that makes a number of high-quality hearing-related devices, including PSAPs. The merged company is one of many to watch during the next few years, as hearing-aid regulations change and as more and more sophisticated tech companies create ear-related products for aging boomers.

Each of my Etymotic plugs is shaped like a Christmas tree: it has three nesting umbrella-shaped flanges made of soft silicone. Earplugs of this type create excellent seals in most people's ears, but inserting them properly can be tricky. The best technique is to widen and slightly straighten your ear canal by reaching over the top of your head with the opposite hand and gently pulling up your pinna. Doing this makes the earplug easier to slide in correctly. (Once it's in, you let go of your ear.) My plugs came with a plastic carrying pouch, which I've attached to my key chain. The first time I used them was at the movie *Dunkirk*, which my wife and I were seeing in an Imax theater. The sound track seemed to consist almost entirely of seat-shaking explosions, but the plugs dialed the bombs back to a tolerable level without making the dialogue impossible to hear. The movie sounded so normal to me that after a while I began to wonder whether the plugs were really doing anything. I pulled one out to check—and discovered that the difference they were making was *huge*. I've also worn my Etymotic plugs on the subway, in both Boston and New York, and while doing loud stuff in my house. I also have a pair of Earos One musician's plugs. They're moderately expensive—$40 a pair—and they are bulkier than my Etymotic plugs, but they fit really well. The right and left plugs (so labeled) are different, and when you insert them correctly, with a little twist, they lock in place.

I own several sets of similar-looking plugs made by Pluggerz, another company with a large line of hearing-protection products, including plugs intended specifically for swimming, shooting, flying, and listening to music. Their least expensive plugs for sleeping—Uni-Fit Sleep, which sell for $15 or $20 a pair—work as well as Flents do but, in my experience, are less likely to come loose during the night. And I find them to be comfortable despite the fact that I sleep on my side, with one ear pressed against the pillow. Pluggerz Sleep plugs comes in one size only, and aren't right for every ear canal. This is true of many hearing-related products—including Apple AirPods, which many people love but others find painful to wear. A good, inexpensive alternative is gumdrop-like silicone earplugs, made by Mack's and others. I used to cut them in half, the better to jam them into my ear canals, but I eventually realized that they work better if you keep them whole and sort of smoosh them over your entire ear opening, as if you were sticking bubble gum to the bottom of a desk.

For several months, I abandoned my Uni-Fit Sleep plugs for Bose Sleepbuds, which were electronic earpieces that played a masking sound that I selected and adjusted on my phone. My usual choice was "Cascade," which was supposedly a waterfall but to me sounded like a window fan. The first night I wore them, I got up to pee and wondered why the fan was following me into the bathroom. They cost $250—a lot for earplugs—but I loved them, and I consistently slept better when I wore them. But then they broke. Bose replaced them, no questions asked. Then they broke again, and then again, and, finally, Bose took them off the market. Sleepbuds had at least two fatal flaws: rechargeable batteries that stopped recharging properly, in some cases after just a few cycles, and tiny magnets that held the earpieces in place in their charger but had a tendency to come unglued. Bose is said to be working on replacements—and I hope they succeed, because I loved my Sleepbuds. One of their best features was an alarm clock that only I could hear, and whose volume I could adjust, making it possible for me to get up early without waking my wife or startling myself.

Nighttime noise has always been a problem for me. Usually when I spend a night in a hotel I acoustically MacGyver my room before going to bed, by silencing everything that can be silenced: I shut down my laptop, unplug the alarm clock, switch the HVAC system either off or to constant fan, unplug the room phone, and shut down or unplug the mini-fridge, to keep the compressor from murdering sleep by repeatedly cycling on and off. Nighttime noise is an especially tricky issue when you travel with grown men, something I do fairly often. On a ski trip when my kids were in high school, I shared a room with two other fiftyish dads and spent much of the night trying to decide which was more annoying, the chain saw–like snoring or the innumerable brief trips to the bathroom. When my friends and I travel for golf and haven't sprung for single rooms, we try to pair the snorers and coughers with the heaviest sleepers, and make sure that the guy with the anti-apnea machine shares a room with someone who likes white noise.

In all the years since I last went to summer camp, I've fired guns just once: in New Zealand, while hunting possums at night with a retired British army commando. (Possums are an existential threat to New Zealand's national bird, the ground-dwelling kiwi, so you are allowed and even encouraged to hunt them at will.) Mostly we used a .22 that had been fitted with a silencer, but we also used a shotgun, which was far from silent. At the time, I didn't give a moment's thought to my ears, but if I ever go hunting again I will wear protection. Hunters and other recreational shooters have many options nowadays, and are less likely than they used to be to scoff at using protection. The Cabela's online catalog offers many pages of electronic devices that amplify quiet sounds, like snapping twigs and whispered conversations, while also muffling gunshots. Devices like those should be considered standard equipment for anyone who uses firearms. If I had just enlisted in the armed forces, I'd buy several kinds and take them with me, just in case.

Friends of mine who know that I've been working on this book have asked me what they ought to do about their own ears. No one should ever take medical advice from a freelance writer, but for someone who has mild or moderate sensorineural hearing loss, I like the approach that Kevin Franck described to me at Mass. Eye and Ear: start with free and work your way up the cost ladder. If you have an iPhone, download the EarMachine app and try it with headphones. If you don't have a smartphone—true of many people in the prime hearing-loss years—the stand-alone SuperEar units that Franck showed me, from Sonic Technology Products, are good devices to start with. Or try a somewhat more expensive PSAP, like the CS50+, from Sound World Solutions, to get a sense of what hearing devices can and can't do for you. Or splurge on Hearphones. Then, if you find that you need or want more, or if it's possible that your hearing problem was caused by something other than ordinary aural wear and tear, make an appointment with a professional. If I suspected that I needed hearing aids and didn't have a pair already sitting unused in a drawer, I'd wait for one of the new, over-the-counter, self-adjustable models, at least one of which should be available by the time you read this. And if I didn't want to wait I'd probably go to Costco, which sells high-quality hearing aids for less than you can buy them in most other places. And I'd load up on noise protection.

THE WORLD SEEMED EERILY QUIET TO ME during the week following 9/11, and at last I understood why: all civilian air traffic in the United States had been halted, and, except for birds and insects, the sky was largely silent. I live in a small town far from any airport, and I had seldom consciously noticed even single airplanes flying overhead. Once all of them were grounded, however, their absence seemed almost palpable, and I realized that I had been hearing them, subconsciously, all along. The sound that airplanes make contributes to the fluctuating din that now ceaselessly hovers just under awareness virtually everywhere in the world. From early October until the snow falls, the dominant sound in my neighborhood is made by gasoline-powered leaf blowers, which drone all autumn long, like aural background radiation. The sound is nearly constant, from seven or eight in the morning until the lawn crews knock off for the day, at four or five in the afternoon. As is the case with most constant noises, I notice it mainly when it starts and when it stops.

Context matters. The first winter my wife and I spent in our house, more than thirty years ago, I could be kept awake by the furnace in the basement—the sound of money burning up—but now the same sound helps me fall asleep, because when I hear it I know there's oil in the tank and our power hasn't gone out. One of my favorite night-time sounds is the high-pitched chirping of the frogs called spring peepers. To me, peepers are the surest sign that winter is finally over, and even when they're at their loudest—their vocalizations have been likened to choruses of sleigh bells—I open the windows in our bedroom as wide as I can, to hear them better as I fall asleep. Yet if exactly the same sound, at exactly the same volume, were being made by a neighbor's car alarm, I'd be on the phone to the police. I once attended a conference in Sausalito, near the northern end of the Golden Gate Bridge. A foghorn in San Francisco Bay sounded at regular intervals throughout the night, and I opened my room's windows wide for it, too. But if I ever heard the same noise at home, and knew it was being made by the guy next door. . . .

"Sound is deeply tied to our emotions in a way that vision is not," Dan Gauger, at Bose, told me. "It's also a sense that we have comparatively little control over. You can look away from something you don't want to see more easily than you can 'hear away' from something you don't want to hear: you can't squinch your ears." Various studies have shown that exposure to low-level noise comparable to that experienced by workers in open-plan office environments can have a variety of deleterious effects, including impacts on health. "It's not the loudness or intensity of noise that makes it stressful for people," Gauger continued. "It's the uncontrollability of it."

Yet gaining even limited control over noise is difficult. I spent twenty years on my town's zoning commission, the last six as the chairman, and we made several attempts at regulating sound, none of them fully satisfactory. "Loudness" is an imprecise standard, and, besides, annoyance is highly subjective. The classic example is the dripping faucet in the bathroom at the end of the hall. That's a noise that many people find intolerable, yet there's no way to describe the nuisance as a number of decibels. You could place a noise meter on your pillow, and angrily stare at it all night, as the dripping kept you awake, yet the sound wouldn't even register on the dial.

A town about an hour from my town has a comprehensive noise ordinance, including limits based on decibel levels. Enforcement is assigned to the police department, which "shall be responsible for investigating, and documenting through acoustic measurements, violations of this ordinance." That means the police not only have to carry sensing equipment, in addition to being given basic training in acoustics, but also have to give a shit. And, as is typical of most such ordinances, most of the noises that are likely to be annoying are specifically exempted: "noises created by snow removal equipment"; "municipal parades, fireworks, historical reenactments, concerts and sporting events"; "noise generated by engine-powered or motor-driven lawn care or maintenance equipment"; "construction equipment while engaged in Premises construction"; "state or municipally authorized and licensed drilling or blasting"; and "solid waste and recycling collection." What's left?

Of course, in order to be bothered by any of this stuff, you have to be able to hear it. Toward the end of my visit to the American School for the Deaf, I got a preview of the school's brand-new Rockwell Visual Communications Center, which was then in the final stages of construction. Its features include a sixteen-foot video screen—flanked by two smaller screens, one for closed captions and one for a sign-language interpreter—and a control room filled with fancy electronics. One regular use for the space will be video conferences with students at schools in other parts of the world. In addition, the room has theater seating, which is mounted on risers that can be pushed back against the rear wall, to create a large open space. There are also ten enormous subwoofers, mounted directly on that floor. Their purpose is to make the building shake sufficiently to enable deaf students to dance to music they can feel but not hear. (Gallaudet University's football team used to snap the ball on a count synchronized with the thudding vibrations from a huge bass drum on the sideline. Gallaudet also invented the football huddle, which made it easier to call plays in sign language.) The subwoofers weren't hooked up yet, but even without them there was plenty of noise in the communications center. Carpenters and electricians were hurrying to finish everything in time for a big opening reception, which at that point was just a week away. They were climbing ladders and screwing down moldings and assembling seats and using electric saws to cut pieces of trim, and they were shouting at each other over the piercing whine of their many power tools. But not one of them, as far as I could see, was wearing hearing protection. What were they thinking?

Acknowledgments and Selected References

This book began as an article for *The New Yorker,* where I had indispensable help from Leo Carey, Henry Finder, Yasmine Al-Sayyad, Dorothy Wickenden, and David Remnick. I learned a great deal from many helpful people not named in the book or in the notes that follow, among them Jeffrey Abelson, Amy Adelman, Irene Allen, Alphonse Altorelli, Mark Beebe, Les Blomberg, Lauren Brown, Bob Carney, Stephen Chao, Roz Chast, Zeng-Yi Chen, James Cramer, Suzanne Day, Leah Driscoll, Erin Duggan, Fern Feldman, Daniel Fink, Laura Friedman, David Girard, Bruce Gordon, Robert Greenspon, Bob Hacker, David Handelman, David Hanscom, Bob Hart, Cliff Henricksen, Christopher Hodgman, Joanna Hodgman, Natalie Howell, Chris Ickler, Ken Jacob, Darlene Ketten, Kathy Krisch, Alex Marshall, David Mason, Lauren McGrath, Seymour Morris Jr., Paula Myers, John Paul Newport, Dan Ostergren, Anne Owen, Carol Owen, Jim Paisley, Daniel Payan, Eugene Pinover, Steve Radlauer, William Self, Charles Shamoon, Mark Singer, Don Steinberg,

Juliette Sterkens, Peter Tyack, Ellis Weiner, Stuart Wolffe, Eric Zwerling, the many people who emailed me descriptions of their own hearing difficulties, and everyone who said "What?" when I told them what I was working on. Thank you also to my agent, David McCormick; my editor, Courtney Young; her assistant, Kevin Murphy; Courtney's boss, Geoff Kloske; my wife, Ann Hodgman; our children, John Bailey Owen and Laura Hazard Owen; and our grandchildren, Alice and Hugh O'Keefe, whose ongoing acquisition of language has demonstrated to me, over and over, the importance of being able to communicate easily with other people.

Google has made many traditional citations superfluous, at least for those of us who aren't trying to earn tenure. The sources listed below are ones that wouldn't necessarily be easy to find by doing a simple Web search based on the text. I've also added some material of possible interest. If I've left out anything important (or made errors), I'll be happy to answer questions (and make corrections) through my website, http://www.davidowen.net/.

ONE. PARDON?

2 **My sister's deafening music:** Here is Anne's Angry and Bitter Breakup Song Playlist. If you play it, don't play it as loud as she did: (1) "Fuck You," Cee Lo Green; (2) "Rolling in the Deep," Adele; (3) "Back to Black," Amy Winehouse; (4) "Stronger (What Doesn't Kill You)," Kelly Clarkson; (5) "Who Knew," P!nk; (6) "You Oughta Know," Alanis Morissette; (7) "Anything but Down," Sheryl Crow; (8) "Harden My Heart," Tiffany; (9) "Raise Your Glass," P!nk; (10) "So What," P!nk; (11) "Bye Bye Bye," NSYNC; (12) "Goodbye to You," Patty Smyth (featuring Scandal); (13) "Goodbye, You Suck," Shiloh; (14) "Wish Me Well (You Can Go to

Hell)," The Bouncing Souls; (15) "Puke," Eminem; (16) "F**kin' Perfect," P!nk; (17) "Without You," *My Fair Lady* Original Broadway Cast.

3 **Hearing-loss statistics:** The same numbers tend to be cited over and over, even year after year, although different authorities cite different ones. Most estimates are probably underestimates, but inaccuracies don't matter very much because all the numbers are incomprehensibly huge. Probably the most a non-statistician needs to know is that hearing loss is common across a range of ages, and is especially common among seniors.

At the same time, there is evidence that, despite a steady increase in the volume level of modern life, Americans hear better than they did fifty years ago. Robert Dobie told me, "This is counterintuitive to some people, but the trends over decades have been that, age for age and sex for sex, we are hearing a little better than our parents and grandparents of the same age and sex. And that's even true in the youngest groups, who one might think were being harmed by the recent increases in recreational music exposure." The explanation, Dobie said, probably has to do with things like the loss of manufacturing jobs in the United States, improvements in ear protection, and the availability of better cardiovascular care. But as some threats to hearing are declining, others are increasing. It's still true that, the older you get, the more people you know who have trouble hearing.

4 **Xerox:** My book about the Xerox machine is *Copies in Seconds* (New York: Simon & Schuster, 2004).

5 **Noisy world:** Human noise pollution doesn't affect just humans: all organisms evolved in a world that was quieter than the one we've created during the past couple of centuries. Human-created sound can be devastating to creatures of all kinds, especially those that depend on hearing for survival—a topic for a book of its own.

6 **Deaf or blind?** Bill Rabinowitz, the head of acoustic research at Bose, introduced in chapter nine, told me that he agreed with Helen Keller on the deaf-or-blind question. "We all know what it's like to be blind: you

just close your eyes, and you think, 'Wow, that's terrible,'" he said. "But most people have no idea what it's like not to hear, because it's difficult to re-create that experience." Earplugs and earmuffs make things quieter, but, if your auditory system is functioning, some sound still reaches your inner ear, both through the muffling devices and through your skull. "To make people temporarily truly unable to hear, you have to put in an attenuating earbud, and then place big hearing protectors over those, and then play some noise through the earbuds, to mask what still gets through," he continued. "I did something like that in a management-training class at the business school at MIT, and we sent people out to walk around and go to lunch. Each one of them had to have a hearing buddy, because if you try walking down the street when you truly can't hear anything, it's dangerous as hell."

8 **Other health effects of hearing loss:** See Katherine Bouton, "Higher Medical Bills for Those Who Don't Treat Hearing Loss," *AARP*, April 19, 2016, https://www.aarp.org/health/conditions-treatments/info-2016/hidden-medical-cost-of-untreated-hearing-loss.html. Bouton herself is hard of hearing, and is the author of the excellent book *Shouting Won't Help* (New York: Sarah Crichton Books, 2013). You can read more on this same subject in Annie N. Simpson, Kit N. Simpson, and Judy R. Dubno, "Higher Health Care Costs in Middle-aged US Adults with Hearing Loss," *JAMA Otolaryngology—Head & Neck Surgery* 142, no. 6 (2016): 607–9, https://jamanetwork.com/journals/jamaotolaryngology/fullarticle/2507066.

9 **Golf, hockey, tennis:** Peter Morrice's article about the effect of hearing on golf is "The Search for Feel," *Golf Digest*, June 2005, page 174. Liam Maguire's experiment with hockey players is described in "Hearing Loss and Sports: How It Affects Performance," on the website of Helix Hearing Care, a Canadian hearing-aid and healthcare provider: https://helixhca.com/general/hearing-loss-and-sports-how-it-affects-performance/. The article about the tennis player is Ben Rothenberg, "For Deaf Tennis Player, Sound Is No Barrier," *New York Times*, November 22, 2016, https://

www.nytimes.com/2016/11/22/sports/tennis/deaf-player-lee-duck-hee
-south-korea.html.

10 **Peggy Ellertsen:** Her 2017 interview is here: "Faces Behind the Screen:
Peggy," 3PlayMedia, January 16, 2018, https://www.3playmedia.com
/resources/faces-behind-screen/peggy/.

TWO. OUR WORLD OF SOUND

14 **Speed of sound:** "Elasticity," as scientists (properly) use the term, means
almost exactly the opposite of what the average person believes it means:
it has to do with how quickly a material returns to its original shape after
some force has been applied to it and then removed. Thus, steel is more
elastic than rubber, because if you place a weight on a piece of steel and
an identical weight on a piece of rubber, then remove both weights, the
steel will return to its original shape faster than the rubber will. Sound
moves more quickly through materials in which the bonds between the
molecules are more elastic. It also moves more quickly through less
dense materials, in which the molecules are less tightly packed.

In pure carbon dioxide (less elastic, more dense), sound moves
about 20 percent more slowly than it does through ordinary air; in
hydrogen (similarly elastic, less dense), it moves almost four times as
fast. It moves through water (more dense but also much more elastic)
two and a half times as fast as it does through air, and it moves through
warm water (less dense) faster than it does through cold, and it moves
through saltwater (more dense but also more elastic) faster than it does
through fresh. It moves through wood (more elastic, less dense) four
times faster than it does through water, and it moves through diamond
(much, much more elastic) at about 12,000 meters per second, or almost 27,000 miles per hour. The speed of light is always the same; the
speed of sound depends.

14 **Andrew Pyzdek,** *Acoustics Today*: The first article in Pyzdek's series,
about the speed of sound, is Andrews "Pi" Pyzdek, "The World Through

Sound: Sound Speed," *Acoustics Today*, n.d., http://acousticstoday.org/the -world-through-sound-sound-speed/. From that article you can follow links to its successors. This is a difficult subject, and Pyzdek is good at explaining it.

16 **Feeling sounds:** People who have what's known as auditory-tactile syn-esthesia, in which particular sounds induce particular sensations in the skin, actually do something like this. Hallucinogenic drugs also have similar effects—or so people say.

17 **The anatomy and function of the ear:** A very clear explanation is by Peter. W. Alberti, of the University of Toronto, "The Anatomy and Physi-ology of the Ear and Hearing," on the website of the World Health Organi-zation, http://www.who.int/occupational_health/publications/noise2.pdf. It's a chapter from the book *Occupational Exposure to Noise: Evaluation, Pre-vention and Control*, published by the World Health Organization in 2001.

21 **Moths and ear mites:** Asher E. Treat, of the City College of New York, published "Unilaterality in Infestations of the Moth Ear Mite" in the *Journal of the New York Entomological Society* 65, no. 1–2 (March–June 1957): 41–50. I read his article in *Insect Lives*, an anthology edited by Erich Hoyt and Ted Schultz (Cambridge, MA: Harvard University Press, 1999).

27 **SoundPrint:** You can read more on the company's website or in Patricia Marx, "Yelp for Noise," *The New Yorker*, October 1, 2018. *Bon Appétit* pub-lished an article about noisy restaurants in 2010: Bridget Moloney, "3 Rea-sons Why Restaurants Are So Loud," *Bon Appétit*, April 20, 2010, https:// www.bonappetit.com/test-kitchen/ingredients/article/3-reasons-why -restaurants-are-so-loud.

28 **Cupping ears with the hands:** Studies have shown that a cupped hand improves the intelligibility of speech and increases the sound level by as much as 10 decibels. Useful devices, hands!

29 **Pre-radar aircraft detectors:** A fascinating account, along with many remarkable photographs, can be found in "Aircraft Detection Before Radar, 1917–1940," Rare Historical Photos, https://rarehistoricalphotos .com/aircraft-detection-radar-1917-1940/.

THREE. THE BODY'S MICROPHONE

31 **Inner ear anatomy and function:** Some excellent animated video explanations show how ears work. Search YouTube for "Auditory Transduction," by Brandon Pletsch, or "Ear Organ of Corti (Full Version)."

34 *"petite madeleine":* Tobias Reichenbach and A. J. Hudspeth, "The Physics of Hearing: Fluid Mechanics and the Active Process of the Inner Ear," *Reports on Progress in Physics* 77 (July 8, 2014).

34 **"the hearing molecule":** See B. Pan et al., "TMC1 Forms the Pore of Mechanosensory Transduction Channels in Vertebrate Inner Ear Hair Cells," *Neuron* 99, no. 4 (August 22, 2018): 736–53.

36 **Absolute pitch:** Familiar music poses a challenge for Komanoff: "When I listen now to Coltrane's 'Spiritual,' I hear it in the C minor key in which it was performed, because I *know*, from fifty-year-old associations, that it is in C. If I were to hear 'Spiritual' today for the first time, though, I would hear it in D minor. Trying to pick out the chords being played by the pianist (McCoy Tyner) would be excruciatingly difficult because of my inability (without heroic effort) of reconciling the notes I would 'hear' in the recording (in D) with the notes I would be forced to play on the piano, in C, in order to duplicate what is actually on the record."

Daniel Levitin and Susan E. Rogers published a fascinating paper, titled "Absolute Pitch: Perception, Coding, and Controversies," in the journal *Trends in Cognitive Sciences* in 2005, when both were at McGill University in Montreal. Levitin wrote me recently: "There's a wonderful example of an improviser being left in the dark—at the beginning of 'Shadows in the Rain' from Sting's first solo album, *Dream of the Blue Turtles*, the drummer, Omar Hakim, has started the tune and apparently the sax player, Branford Marsalis, doesn't know what key they're in. You can hear him ask, 'What key is it in? Wait! Wait! What key is it in?' I don't know what happened next, but from having played with Sting myself I can testify that this is a common experience. I imagine that Branford either played quietly to himself while Sting was singing, to determine

the key, or he looked at the bass player's or piano player's fingers to figure it out. If he had absolute pitch, he wouldn't have to do that. But he
also wouldn't be able to as deftly navigate from one key to another on
the fly, a mark of good relative pitch." (Levitin, after emailing me, asked
Marsalis himself how he had determined the key of that song, and Marsalis confirmed that he had used Sting's vocal.)

37 **Losing absolute pitch:** A good overview is Mary L. Bianco, "Understanding and Dealing with the Loss of Absolute Pitch as One Ages"
(master's thesis, Mills College, 2015), https://tinyurl.com/ybendjf2.

You can test your own relationship with pitch by taking an online
version of a test offered by the National Institute on Deafness and Other
Communication Disorders, "Test Your Sense of Pitch," NIDCD, July 31,
2014, https://www.nidcd.nih.gov/tunestest/test-your-sense-pitch.

38 **Evolution of balance and hearing:** David Corey: "It's all speculation,
but the hair cells in the balance organs are more like hair cells in lower
vertebrates, in their shape and in the way the organs grow. . . . A frog,
for instance, has a hearing organ, and throughout its life it's adding
more and more cells at the edge of the hearing organ. And . . . mice,
in the vestibular part . . . tend to continue to add . . . some hair cells,
at least for the first month or two. So . . . the kind of default thing is
to be always growing a few more cells and a few more cells."

40 **Vestibular catastrophe:** John Crawford's account of his loss of his sense
of balance is J.C., "Living Without a Balancing Mechanism," *New England Journal of Medicine* 246, no. 12 (March 20, 1952): 458–60.

FOUR. WHEN HEARING FAILS

47 **Decibels:** I asked Erich Thalheimer, a technical specialist for acoustics
and vibration at WSP USA, an international engineering consulting firm,
if he could explain decibels, sound pressure, sound intensity, and sound
energy in a single, layman-friendly paragraph. He wrote, in an email:

"Sound is simply a variation or vibration in air pressure that we can detect with our auditory system (i.e. hear). The decibel scale, named after Alexander Graham Bell, is a logarithmic scale adopted by acousticians for simplicity so that the extremely large range of air pressure fluctuations that we can hear could be expressed more easily. Sound can be described and measured in many different ways. A source will emit what is known as a sound power level (Lw) which is a property of the source itself completely independent of its environment. As the sound propagates away from the source it does so as a sound intensity level (Li), or the amount of sound power per unit area traveling in a given direction. As the sound reaches a receiver (i.e. listener), we can hear it as a fluctuation in air pressure, or a sound pressure level (Lp), based on how the pressure waves have interacted with and been affected by the environment."

51 **Dysecoea:** You can find an 1800 edition of Cullen's book (the original version of which he wrote in Latin) on Google Books, by searching for "William Cullen *Nosology*."

51 **Andrew Ferguson's letter about blacksmiths:** You can read a facsimile here: https://www.ncbi.nlm.nih.gov/pmc/articles/PMC5668781/pdf/medphysj68552-0040.pdf.

52 **Boilermakers and others:** Thomas Barr's "Enquiry into the Effects of Loud Sounds upon the Hearing of Boilermakers and Others Who Work amid Noisy Surroundings," published by the Royal Philosophical Society of Glasgow, can be downloaded here: https://archive.org/details/b21457384. Thomas Oliver's book is *Dangerous Trades* (London: John Murray, 1902). C. C. Bunch's article is "Conservation of Hearing in Industry," *Journal of the American Medical Association* 118, no. 8 (February 21, 1942): 588–93, https://jamanetwork.com/journals/jama/article-abstract/253913?redirect=true.

56 **The brave Confederate colonel:** You can read a full account here: "Hearing Loss After a Battle," CivilWarTalk, October 13, 2013, https://civilwartalk.com/threads/hearing-loss-after-a-battle.90844/page-2.

57 **Arthur Cheatle and "gun deafness":** Arthur H. Cheatle, "Gun Deafness and Its Prevention," *Royal United Service Institution Journal* 51 (1907),
available on Google Books, https://tinyurl.com/y8h24ubp.

61 **Military study of the effectiveness of the V-51R earplug:** You can read
the full report: Bernard Jacobson, Elizabeth M. Dyer, and Robert J. Ma-
rone, "Effectiveness of the V-51R Ear Plug with Impulse Pressures up to 8
psi," Human Engineering Laboratories, Aberdeen Proving Ground, Maryland, November 1962, https://apps.dtic.mil/dtic/tr/fulltext/u2/401212.pdf.

62 **Stephen Carlson's article about the effect of military service on
ears:** "We Treat Hearing Loss as an Inevitable Cost of War. It Shouldn't
Be," *Washington Post*, April 12, 2016, https://tinyurl.com/ybkrtcux.

FIVE. CICADAS IN MY HEAD

70 **What tinnitus sounds like:** There are many websites that let you listen
to audio files that simulate some of the many forms that tinnitus can
take. Here's one: Sound Relief Hearing Center, "Sounds of Tinnitus,"
n.d., https://www.soundrelief.com/tinnitus/sounds-tinnitus/.

83 **John Shea, *Life* magazine:** The issue—September 14, 1962—is from
the golden age of magazines (check out the number of advertisements).
It's available on Google Books at https://tinyurl.com/y9b7njcg.

86 **"Reddit Tinnitus Cure":** Charles Liberman: "Many people can modu-
late their tinnitus by manipulations that stimulate the somatosensory
system in the head and neck region. The underlying idea is that many
auditory regions in the brain also get input from the somatosensory region, because for your brain to correctly compute the locations of sounds
in space, it helps for it to know if your head is turned, or if your pinna
is cocked one way or the other (for animals with movable pinnae). This
general line of thinking also explains why some people can get tinnitus
simply by non-ear-related injury to the head and neck region."

And addressing that injury can sometimes eliminate the tinnitus or
reduce its severity. A reader told me in an email: "I developed high-pitched

tinnitus in one ear. It changes when I turn my head and gets worse when I lie down. I met with a skilled physical therapist, who examined my neck and determined that my tinnitus is a direct relation to my upper cervical spine resulting from poor posture (I work at a computer). The treatment is physical therapy." I asked if the therapy had made her tinnitus go away. "It has helped a lot, for sure," she wrote. "It's not completely gone, but the two physical therapists I've been working with tell me they do expect it to go away completely. The neck exercises I do at home really seem to help, along with working on my posture and sleeping on one pillow. The test that you can do on yourself is to turn your head in both directions (as if looking over your shoulder) to see if your tinnitus changes at all. I believe that people who have suffered from tinnitus for years may not realize that the neck could be the culprit."

87 **Phantom limbs:** John Colapinto, "Brain Games," *The New Yorker*, May 11, 2009, https://www.newyorker.com/magazine/2009/05/11/brain-games; and Atul Gawande, "The Itch," *The New Yorker*, June 30, 2008, https://www.newyorker.com/magazine/2008/06/30/the-itch.

89 **Desyncra:** The paper about research sponsored by the company can be found here: Christian Hauptmann et al., "Technical Feasibility of Acoustic Coordinated Reset Therapy for Tinnitus Delivered via Hearing Aids: A Case Study," *Case Reports in Otolaryngology* (2017), https://www.ncbi.nlm.nih.gov/pmc/articles/PMC5390560/. General Fuzz's free version of the same idea is here: "ACRN Tinnitus Protocol," http://generalfuzz.net/acrn/. A related treatment is tinnitus retraining therapy, whose leading developer and proponent has been Pawel J. Jastreboff: https://www.ncbi.nlm.nih.gov/pubmed/25862626.

90 **General Fuzz:** A reader told me in an email: "I have had thirty years or more of various internal sounds that have been diagnosed as tinnitus. A friend taught me that, if I could match the sound in my head by humming, it would fool my brain into stopping the noise—sometimes for moments (and I'd have to repeat the humming) but often for days or even weeks. Having control over the sound is a huge relief and so simple.

However, over time the sound in my head has shifted from a high-pitched, hummable noise to a more mechanical noise. I was distressed that I could not make this sound in my throat, but then I discovered that I could download a white-noise app to my phone which can be adjusted to match my new sound range."

Another reader, a musician, told me that his tinnitus went away entirely after he took voice-training lessons for opera. He wrote, "I strongly suspect that learning how to vocalize more or less properly, with emphasis on developing natural vibrato which occurs in your head and nasal cavities, must have reduced the tinnitus so gradually that I didn't notice when it was gone."

SIX. CONDUCTIVE HEARING LOSS

110 **Bone conduction:** Scientists at the Naval Submarine Medical Research Lab, in Groton, Connecticut, have found that when humans are underwater they can detect sounds well into the ultrasound region—in fact, at frequencies higher than 200,000 hertz, which is at the upper end of the porpoise range. Those sounds are conveyed to the cochlea not by way of the eardrum but through bone conduction, which may be the main mechanism by which people hear underwater. Michael K. Qin et al., "Human Underwater and Bone Conduction Hearing in the Sonic and Ultrasonic Range," *Journal of the Acoustical Society of America* 129, no. 2485 (April 8, 2011), https://asa.scitation.org/doi/10.1121/1.3588185.

113 **Lempert's fenestration:** The YouTube web address is https://www .youtube.com/watch?v=8YJ44qw61O0.

SEVEN. HEARING AIDS

122 **"Deaf flights":** More on this bad idea, from *Smithsonian*: Greg Daugherty, "Doctors Once Prescribed Terrifying Plan Flights to 'Cure' Deafness," Smithsonian.com, September 26, 2017, https://www.smithsonianmag

.com/history/doctors-once-prescribed-terrifying-plane-flights-cure -deafness-180965027/.

123 **Acoustic throne:** You can see a picture of the king of Portugal's acoustic throne here: Robert Traynor, "Joao's Acoustic Throne," Hearing Health and Technology Matters, July 28, 2015, https://hearinghealthmatters.org /hearinginternational/2015/joaos-acoustic-throne/.

127 **Home hearing test:** You can take a home version of a standard hearing test over the telephone, by going here: "The National Hearing Test," https://www.nationalhearingtest.org/wordpress/?page_id=2730. The test is free for AARP members, $8 for everyone else. From the website: "You listen to three-digit sequences presented in a background of white noise and then enter the digits using the telephone keypad. Similar tests have been used with great success throughout Europe and in Australia; this is the first of its kind in the U.S. It was developed with funding from the National Institutes of Health and is provided on a nonprofit basis, although there is a small fee for it. The goal is to give you information that can help you decide whether you should seek a full-scale evaluation of your hearing." Another home hearing test—on a site that has much else to offer—is at https://www.mimi.io/mimi-hearing-test.

130 **Tinnitus masker in hearing aids:** The two masking channels in my hearing aids—which my audiologist selected based on what she had deduced about my tinnitus from the way I described it during my exam— are not controllable by me. I can play either channel, or neither, but I can't adjust their volume, and I can't replace them with other masking sounds without going back to an audiologist. The impact of tinnitus is subjective. Sometimes I notice mine a lot, and sometimes I don't notice it at all, and how loud it seems to me at any moment depends on what I'm doing and thinking about, and what's going on around me. You can't effectively mask something like that with a pair of preset sounds selected by someone else.

131 **Reddit discussion about hearing aids and cochlear implants:** "No Longer Deaf People of Reddit What's Something You Thought Would

Have a Certain Noise but Were Surprised It Doesn't?" reddit, https://www.reddit.com/r/AskReddit/comments/9wdvtk/no_longer_deaf_people_of_reddit_whats_something.

151 **Google Translate:** An excellent discussion of the limitations of this (nevertheless often quite useful) utility, by the author of *Gödel, Escher, Bach*, is Douglas Hofstadter, "The Shallowness of Google Translate," *The Atlantic*, January 30, 2018, https://www.theatlantic.com/technology/archive/2018/01/the-shallowness-of-google-translate/551570/.

153 **Starkey trial:** The indictment: The United States Attorney's Office, District of Minnesota, "Five Indicted for Massive Fraud Perpetrated Against Starkey Laboratories," September 21, 2016, https://www.justice.gov/usao-mn/pr/five-indicted-massive-fraud-perpetrated-against-starkey-laboratories. A good account of Starkey's legal difficulties is Michela Tindera, "Runaway Billionaire: Meet the CEO Whose Company Descended into Fraud, Embezzlement and Betrayal," *Forbes*, June 30, 2018, https://www.forbes.com/sites/michelatindera/2018/06/12/starkey-hearing-bill-austin/#57e0489c3090.

EIGHT. STIGMA

157 **Charlie Rose hearing-loss show:** The show aired on October 11, 2013. You can watch it at https://charlierose.com/videos/17843.

158 **Alice Cogswell:** I learned about Alice Cogswell from Gary Wait, who retired as the archivist of the American School for the Deaf in 2013 but still helps out. There's more about Wait and ASD here: CTMQ, "103. American School for the Deaf Museum," http://www.ctmq.org/103-american-school-for-the-deaf-museum/.

160 **Samuel Johnson:** Dr. Johnson's book about the trip that he and James Boswell took is *A Journey to the Western Isles of Scotland*, first published in 1775. You can read the full text at https://www.gutenberg.org/files/2064/2064-h/2064-h.htm.

161 **History and culture of signing:** A good book on this subject is Gerald Shea, *The Language of Light* (New Haven: Yale University Press, 2017). Shea is the deaf lawyer who didn't know he was deaf until he was tested in his thirties; he also wrote *Music Without Words*, quoted in chapter six. In *The Language of Light*, he describes the cumbersome "methodical" system that Laurent Clerc had been taught in France: "The cases of the nouns of French grammar . . . were taught by rolling the right index finger around the left, descending the hands from the first roll (the nominative case), to the second (genitive), to the third (dative), and so forth. The numerous articles in French (*le, la, les, du, des*) were 'signed' by indicating joints of the fingers, wrists, elbows, and shoulders, which, in a simile that had nothing to do with a signed language, showed how articles were *adjoined* to nouns just as appendages were attached to people. . . . To sign the phrase *to look up with extreme pleasure* . . . required thirteen or fourteen methodical signs, including an article and two prepositions" (p. 28).

162 **Chilmark, Massachusetts:** The newspaper article I found in the old scrapbook is Ethel Armes, "Deaf-Mute Community: Chilmark, Martha's Vineyard, Uses Exclusively the Sign Language" (date unknown).

167 **Alexander Graham Bell:** In his 1883 presentation to the National Academy of Sciences, Bell included tables listing the students enrolled at the American School for the Deaf between 1817 and 1877, and argued that the recurrence of certain surnames supported his contention that the deaf were irresistibly and tragically attracted to one another. More than a few of the names belonged to children from Martha's Vineyard. Bell's complete presentation can be read here: Alexander Graham Bell, "Upon the Formation of a Deaf Variety of the Human Race," paper presented to the National Academy of Sciences, November 13, 1883, https://ia800702.us.archive.org/21/items/gu_memoirformati00bell /gu_memoirformati00bell.pdf.

In *Forbidden Signs* (Chicago: University of Chicago Press, 1996), Douglas

Baynton writes that Bell "traveled the country delivering speeches on the dangers of deaf interbreeding, such as the one he gave to the Chicago Board of Education in which he warned that deaf people, 'by their constant association with each other, form a class of society and intermarry without regard to the laws of heredity.'" Bell's arguments, along with similar ones made by other Americans, were used to justify compulsory-sterilization laws in this country, and those laws, in turn, were cited by the Nazis, who in the 1930s and 1940s sterilized thousands of deaf people, including children who had been reported to the authorities by their teachers.

169 **Recording ASL without YouTube:** There are systems for representing ASL symbolically, on a printed page, but the number of symbols and symbol elements is necessarily huge, and, even with a computer program that draws on a vast Unicode symbol set, it can be cumbersome to use. The most widely used is Sutton SignWriting, which was introduced in the 1970s by a dancer who, two years earlier, had created a system for symbolically recording choreography.

169 **More on signing and language from Oliver Sacks:** "Signers with right hemisphere strokes, in contrast, may have severe spatial disorganization, an inability to appreciate perspective, and sometimes neglect of the left side of space—but are not aphasic and retain perfect signing ability despite their severe visual-spatial deficits. Thus signers show the same cerebral lateralization as speakers, even though their language is entirely visual-spatial in nature (and as such might be expected to be processed in the right hemisphere)."

NINE. BEYOND CONVENTIONAL HEARING AIDS

171 *Harper's* **article about funerals:** David Owen, "Rest in Pieces," *Harper's*, June 1983, https://www.davidowen.net/files/rest-in-pieces-6-83.pdf.

172 **The FDA and hearing aids:** The government's official definition of a medical device can be found at U.S. Food and Drug Administration, "Regulatory Requirements for Hearing Aid Devices and Personal

Sound Amplification Products—Draft Guidance for Industry and Food and Drug Administration Staff," November 7, 2013, https://www .fda.gov/MedicalDevices/ucm373461.htm. FDA guidance on hearing aids vs. PSAPs can be found at U.S. Food and Drug Administration, "Regulatory Requirements for Hearing Aid Devices and Personal Sound Amplification Products—Guidance for Industry and FDA Staff," February 25, 2009, https://www.fda.gov/downloads/medicaldevices/devicere gulationandguidance/guidancedocuments/ucm127091.pdf. FDA's approval of Bose's request to sell an OTC hearing aid is U.S. Food and Drug Administration, Letter to Bose Corporation, c/o Deborah Arthur, Re: DEN180026, Trade/Device Name: Bose Hearing Aid, October 5, 2018, https://www .accessdata.fda.gov/cdrh_docs/pdf18/DEN180026.pdf, and U.S. Food and Drug Administration, "FDA Allows Marketing of First Self-Fitting Hearing Aid Controlled by the User," FDA news release, October 5, 2018, https://www.fda.gov/NewsEvents/Newsroom/PressAnnouncements /ucm622692.htm.

TEN. COCHLEAR IMPLANTS

206 **Rush Limbaugh's cochlear implants:** "Where I've Been the Last Week," *The Rush Limbaugh Show*, April 24, 2014, https://tinyurl.com/ycunc5no.

209 **Peggy Ellertsen's listening therapist:** Ellertsen worked hard for years at getting the most out of the hearing she still had, and for some time had spent two hours a week working with Geoff Plant, the founder and director of the Hearing Rehabilitation Foundation (http://hearingrehab .org/). Plant is technically retired, but he still trains and counsels people who have hearing loss, and he had helped Ellertsen boost the effectiveness of her hearing aids by learning to "laser focus" on what people around her were saying. She told me, "You can't do it indefinitely, because it does you in; you have to go home and take a nap. But it's important for people with hearing loss to practice doing it."

Ellertsen had also taken advantage of an assistive device sold by

Phonak, the manufacturer of her hearing aid. That device is the Roger Pen, a rechargeable microphone and Bluetooth transmitter, which can be paired with a hearing aid or cochlear implant, then positioned so that it picks up sound directly from a particular speaker or other sound source. Ellertsen had recently expanded her Roger collection to seven, at roughly $800 per Pen, so that she could fully participate in conversations during a family vacation with half a dozen relatives.

212 **Meaghan Reed's ticking clock:** Peter W. Alberti, "The Anatomy and Physiology of the Ear and Hearing," chapter 2 of *Occupational Exposure to Noise: Evaluation, Prevention and Control*, ed. Berenice Goelzer, Colin H. Hansen, and Gustav A. Sehrndt (Geneva: World Health Organization, 2001), 53–62 (https://www.who.int/occupational_health/publications/noise2.pdf), explains why she couldn't hear it: "Hearing has an alerting function especially to warning signals of all kinds. There are brain cells which respond only to the onset of a sound and others which respond only to the switching off of the sound, i.e. a change. Think only of being in an air-conditioned room when the air conditioner turns on, one notices it. After a while it blends into the background and is ignored. When it switches off, again one notices it for a short time and then too the absence of sound blends into the background. These cells allow the ear to respond to acoustic change—one adjusts to constant sound—change is immediately noticeable. This is true too with machinery and a trained ear notices change."

218 **Juliet Corwin's** *Washington Post* **op-ed:** Juliet Corwin, "The Lonely World Between the Hearing and the Deaf," *Washington Post,* July 20, 2018, https://tinyurl.com/ya8aao6t.

There are those who would argue that I, as a hearing person, have no right to form an opinion about any of this. But shunning by the Deaf of people with implants doesn't seem entirely different to me from the efforts made by people like Alexander Graham Bell to force the deaf to conform to and accommodate the hearing. In both cases, the guiding impulse is a belief that cultural identity is more important than access to what Helen Keller correctly identified as the thing that "sets thoughts astir and keeps us in the intellectual company of men." Human communication, regardless of the form, should be the priority.

ELEVEN. ASYLUM

224 **Andrew Solomon's *New York Times Magazine* article:** Andrew Solomon, "Defiantly Deaf," *New York Times Magazine*, August 28, 1994, https://www.nytimes.com/1994/08/28/magazine/defiantly-deaf.html.

227 **PACES, ASD's program for students who have more issues than deafness:** Karen Wilson, the program's director, told me, "I'll give you an example that's probably the most powerful one we have currently. We got a referral for a little girl who had been at several schools for the deaf in New York. Nobody had been able to manage her, so three of us drove down to see her, at a facility in Yonkers. The place looked like a one-story nursing home. The sign identified it as a school for 'the retarded,' and we thought, This isn't going to be good—they're not even using appropriate language." The girl was eleven or twelve years old, and she was wearing diapers, and she had a large, permanent, scabbed-over wound on her forehead, from compulsively banging her head.

"She was profoundly deaf, but nobody there signed, and she had no language," Wilson continued. "Her room was literally padded, with what looked like gym mats on the walls, and her legs were so weak, I assume from never being exercised, that she moved almost like a mermaid—she just sort of crawled around. I was in the backseat as we drove home, and I said, 'We've got to take her.' She wasn't really appropriate for us, but we couldn't leave her there. The administration agreed, and we told New York that we'd like to give her a nine-week trial."

That was five years ago. "She was out of diapers, just like that," Wilson said. "She didn't need them; she just needed someone to help her do what she needed to do. She showers independently, she dresses independently, she engages in activities. Her receptive skills for sign language are very good, and her sign vocabulary is probably huge. Her expressive skills are less so—she signs simple one- and two-word phrases—but she's intellectually disabled so she's not going to just start signing a lot of content. She's a remarkable kid, though, and there's no comparison between her quality of life then and her quality of life now. She'll be a great candidate for a group home."

A key to her transformation, in addition to finally having caregivers who were qualified to care for her, was having access to language—even at the low level at which she's been able to acquire it. Her case is extreme, but deaf children who grow into school age without an immersive exposure to any language are at an immense, permanent disadvantage, too, and not just in terms of communication skills narrowly defined.

228 **ASD's strategic plan:** American School for the Deaf, *Strategic Plan 2015–2018*, https://www.asd-1817.org/uploaded/Executive_Director/Strategic_Plan_2015-2018.pdf.

TWELVE. THE MICE IN THE TANK

237 **Edwin Rubel:** Proof that we live in a small world: Kevin Franck, now at Mass. General, worked Rubel's lab when he was a graduate student. And Franck and I eventually learned—by accident, through my brother—that his daughter and my niece row on the same college team.

242 **Neurofibromatosis type 2:** Bradley Welling told me, "Most people with NF2 notice tinnitus or hearing loss as the first symptom, but if you interview them carefully you find that about fifty percent of them also have some vestibular issues. NF2 also often results in facial-nerve paralysis, which can affect eyesight, since when your face is paralyzed you can't blink. If you have no hearing, and you have no facial expression, and your vision is impaired—those losses take you out of the realm of communication. The tumors are not considered malignant, but they grow in an area that compresses the brain stem, so they're very damaging, and they can be fatal."

246 **Two landmark papers by Sharon Kujawa and Charles Liberman:** Sharon G. Kujawa and M. Charles Liberman, "Acceleration of Age-Related Hearing Loss by Early Noise Exposure: Evidence of a Misspent Youth," *Journal of Neuroscience* 26, no. 7 (February 15, 2006): 2115–23, http://www.jneurosci.org/content/26/7/2115.long; and Sharon G. Kujawa and M. Charles Liberman, "Adding Insult to Injury: Cochlear Nerve Degeneration After 'Temporary' Noise-Induced Hearing Loss," *Journal of Neuroscience* 29, no. 45 (November 11, 2009): 14077–85, http://www.jneuro sci.org/content/29/45/14077.

246 **Restoration of hearing in mice with induced cochlear synaptopathy:** See Jun Suzuki, Gabriel Corfas, and M. Charles Liberman, "Round-Window Delivery of Neurotrophin 3 Regenerates Cochlear Synapses After Acoustic Overexposure," *Scientific Reports*, 6 (April 25, 2016), https:// www.ncbi.nlm.nih.gov/pubmed/27108594.

247 **Stéphane Maison:** When I met him, Maison was developing methods of diagnosing cochlear synaptopathy that don't involve dissecting the heads of sufferers and examining their inner ears under microscopes. The ability to make accurate diagnoses, he told me, is a critical first step both in understanding the nature of the problem in humans and in testing and implementing a possible cure, should one be found. "We've tried everything we could think of when it comes to measuring hearing thresholds, looking at the middle ear, assessing the reflex in the brain stem," he said. "We do a lot of speech testing, in quiet but also in very challenging environments, and then we look at the nerve response." The hope is that some relatively simple combination of such tests, or perhaps a new test, will prove to be definitive. And, he said, the need is urgent. "The number one complaint of people who get hearing aids is not that that they can't hear; it's that they can't understand," he continued. "Why? Because a hearing aid is a booster of sounds. It replaces the function of the outer hair cells. It doesn't restore connections between the inner hair cells and the nerve fibers that are responsible for intelligibility. It does nothing to address synaptopathy."

248 **"Hidden hearing loss" and cochlear implants:** Implants work even with nonfunctional synapses, because the implanted electrodes directly stimulate the auditory nerve fibers: the signal is strong enough to jump the gap. But in people whose synapses or hair cells were damaged so long ago that the disconnected fibers have receded too far from the cochlea, an implant may no longer be able to fully "plug in."

THIRTEEN. VOLUME CONTROL

256 **Kennels and animal shelters:** Barking causes more hearing problems than you might guess: Chandran Achutan and Randy L. Tubbs, NIOSH Health Hazard Evaluation Report, HETA #2006-0212-3035, Kenton County Animal Shelter, Covington, Kentucky, National Institute for Occupational Safety and Health, February 2007, https://www.cdc.gov/niosh/hhe/reports/pdfs/2006-0212-3035.pdf.

262 **How to insert a push-in earplug:** You can watch Elliott Berger, of 3M, do it on YouTube. Search for "Fitting Push-In Earplugs."

266 **Gasoline-powered leaf blowers:** My old *Atlantic* colleague James Fallows has been active in fighting the leaf-blower scourge in Washington, D.C., where he lives. Not long ago, he wrote to me: "We commissioned research here in DC that showed that these two-stroke gas engines really have a different sound signature [from] other devices whose decibel ratings are 'the same.' Main reason: all the low-frequency noise they generate, which has far greater penetrating power than high-frequency noise." You can read more on the website of Quiet Clean D.C.

267 **Zoning regulations:** A further challenge with attempts to regulate noise is that sound transmission is affected by topography. Valleys can amplify sounds, as amphitheaters do—so much so that a carelessly situated piece of machinery can sound louder to a person a couple of hundred yards away than it does to a person much closer. People often assume that they can neutralize noise by planting a few trees between the source and themselves, but trees, even if you plant a forest's worth, can be

virtually transparent to sound waves, which, unlike light waves, don't travel in straight lines and can easily flow around obstacles. Tall, solid barriers, like the ones you increasingly see alongside highways that transect residential areas, are effective at reflecting traffic noise away from houses directly behind them, but reflected sound doesn't disappear, and if a barrier is built on just one side of a road it makes life worse for people living on the other, by adding reflected road noise to the road noise they were hearing already. In addition, according to a 2017 article by Meryl Davids Landau, in the nonprofit digital magazine *Undark*, "Those living up on hills or near freeway openings sometimes find the noise actually worsens once walls are built nearby." Wind and weather are factors as well: "In the early morning, if the ground is cool but the air warms up, for instance, sound that would normally be pushed up is refracted downward, causing homes some 500 or 1,000 feet from the road to hear it loudly." Meryl Davids Landau, "On Highway Noise Barriers, the Science Is Mixed: Are There Alternatives?" *Undark*, December 27, 2017, https://undark.org/article/highway-noise-barrier-science/.

New York City has a fourteen-thousand-word noise ordinance, the stated purpose of which is "to reduce the ambient sound level in the city, so as to preserve, protect and promote the public health, safety and welfare, and the peace and quiet of the inhabitants of the city, prevent injury to human, plant and animal life and property, foster the convenience and comfort of its inhabitants, and facilitate the enjoyment of the natural attractions of the city." Worthy goals! But in a city where diesel garbage trucks operate all night, and taxi drivers lean on their horns even when they can see that the cars for blocks ahead of them have nowhere to go, the rules can't possibly mean much.

Index

tinnitus (*cont.*)
  finger snapping at base of neck to
    treat, 83
  hearing aids and, 85
  hearing loss accompanying, 69–70
  lidocaine infusions for, 83, 87
  neuromodulation therapy, 89–90
  pain connection, 87–88
  patient reactions to, 72–75, 81–84
  phantom limb pain analogy for, 87–88
  pulsatile tinnitus, 72
  real sound, masking with, 85–86
  Starkey Relax smartphone masking
    app, 141
  sudden hearing loss and, 78
  tumors and, 70–71
T-Mic, 215–16
tobramycin, 44
Total Communication, 223
Tousard, Louis de, 55–56
"Try the McGurk Effect"
  (BBC story), 146

ultrasonic frequencies, 15–16, 35
Uni-Fit Sleep, 263
United Healthcare, 173–74
USS *Constitution* (*Old Ironsides*), 54

Vactuphone, 124
Valsalva maneuver, 18
van Bergen, Leo, 59–60
Van Tasell, Diane, 134–35, 172–75, 177,
  179, 180–82, 185, 201
vestibular rehabilitation therapy, 42
vestibular system, 38–42, 43
vestibule, of ear, 17
V-51R Ear Wardens, 61
viral vectors, 237

visual clues, and hearing, 145–46
voices, pitch of, 128–29

warfare and weaponry, hearing loss
  from, 54–65
  Civil War and, 56–57
  Combat Arms Earplugs, 63–64
  earplugs, use of, 60–62, 63–64
  Industrial Revolution and, 52–54
  Iraq and Afghanistan wars and, 62–63
  naval combatants and, 54–55, 57–59
  Tactical Communications and
    Protective Systems (TCAPS)
    electronic headset, 64–65
  tools invention and, 51–52
  V-51R Ear Wardens and, 61
  World Wars I and II, 59–61
Warren, Elizabeth, 200
Watson (IBM technology), 148
Wawrzonek, John, 74–75, 80,
  99–102, 104–6
Welling, D. Bradley, 241–43
Western Electric, 124
Widex, 125
Williams-Steiger Occupational Safety
  and Health Act, 253
Wilson, Karen, 229
workplace safety
  jobs not subject to OSHA, 255–56
  OSHA regulations, 253–55
World War I, 59–60
World War II, 61
Wynn, Collette, 111–12, 113

Young Thomas Edison (film), 109

Zamir, Lee, 183, 184
Zenith, 124